WEDDED TO MY SWORD

The Revolutionary War Service of Light Horse Harry Lee

Michael Cecere

HERITAGE BOOKS
2012

HERITAGE BOOKS
AN IMPRINT OF HERITAGE BOOKS, INC.

Books, CDs, and more—Worldwide

For our listing of thousands of titles see our website
at
www.HeritageBooks.com

Published 2012 by
HERITAGE BOOKS, INC.
Publishing Division
100 Railroad Ave. #104
Westminster, Maryland 21157

Copyright © 2012 Michael Cecere

Other Heritage Books by the author:

Captain Thomas Posey and the 7th Virginia Regiment

An Officer of Very Extraordinary Merit: Charles Porterfield and the American War for Independence, 1775–1780

Great Things Are Expected from the Virginians: Virginia in the American Revolution

In This Time of Extreme Danger: Northern Virginia in the American Revolution

They Are Indeed a Very Useful Corps: American Riflemen in the Revolutionary War

They Behaved Like Soldiers: Captain John Chilton and the Third Virginia Regiment, 1775–1778

To Hazard Our Own Security: Maine's Role in the American Revolution

Cover painting: "Light Horse Harry Lee" by Pamela Patrick White

All rights reserved. No part of this book may be reproduced or transmitted in any form or by any means, electronic or mechanical, including photocopying, recording or by any information storage and retrieval system without written permission from the author, except for the inclusion of brief quotations in a review.

International Standard Book Numbers
Paperbound: 978-0-7884-5391-5
Clothbound: 978-0-7884-9483-3

Contents

Introduction 1

Ch. 1   "A great spirit of Liberty actuates every Individual" 9

Ch. 2   "He is so enterprising and useful an officer" : 1777 29

Ch. 3   "Captn Lee's Vigilance baffled the Enemy's designs" : 1778 65

Ch. 4   "Capt. Lee's genius particularly adapts him to [this] command" : 1778 79

Ch. 5   "Major Lee has performed a most gallant affair" : 1779 95

Ch. 6   "Major Lee is...a man of great spirit and enterprise" : 1780 123

Ch. 7   "Heavy rains, deep creeks, bad roads, poor horses, and...want of provisions..." : 1781 147

Ch. 8	"Unless our Army is greatly reinforced I see nothing to prevent [the enemy's] future progress" : 1781 175
Ch. 9	"Our poor Fellows are worne out with fatigue" : 1781 207
Ch. 10	"Few Officers...are held in so high a point of estimation as you are" 1781-82 239

Epilogue 267

Bibliography...273

Index..291

Acknowledgements

As with all of my books, I find myself indebted to the Simpson Library at the University of Mary Washington for their extensive collection of resources on the American Revolution. Their resources were tremendously helpful in my research. The staff and resources at the Virginia Historical Society and the Virginia State Library were also extremely helpful as was the research library at Stratford Hall. Leesylvania State Park, the birthplace of Light Horse Harry Lee, was generous in their support of my efforts and I sincerely thank the staff for their assistance. The same goes for the staff at the RELIC room of the Prince William County Library.

My interest and knowledge of the American Revolution has been greatly enhanced by my decade long involvement with Revolutionary War reenacting. Many fellow reenactors and Rev War researchers and enthusiasts contributed directly and indirectly to my efforts with valuable comments and responses to my questions. I am truly grateful to all of them, particularly Patrick O'Kelly for allowing me to use some of his maps from his writings on the Revolutionary War in the Carolinas, Pamela Patrick White for allowing me to use her wonderful artwork, and Bert Dunkerly for sharing his knowledge and research on the Revolutionary War in South Carolina.

As always, I want to thank my wife Susan, and my children, Jenny and Michael, for allowing me to devote as much time as I do to my research, writing, and reenacting.

About the Author

Michael Cecere Sr. is the proud father of Jennifer and Michael Jr. and the grateful husband of Susan Cecere. He teaches American History at Robert E. Lee High School in Fairfax County, Virginia and was named the 2005 Outstanding Teacher of the Year by the Virginia Society of the Sons of the American Revolution. Mr. Cecere also teaches American History at Northern Virginia Community College. He holds two Master of Arts degrees from the University of Akron, one in History and the other in Political Science. An avid Revolutionary War re-enactor, he currently is the commander of the 7th Virginia Regiment and participates in numerous living history events throughout the country. This is his eighth book.

Other books by Michael Cecere

They Behaved Like Soldiers: Captain John Chilton and the Third Virginia Regiment, 1776-1778 (2004)

An Officer of Very Extraordinary Merit: Charles Porterfield and the American War for Independence, 1775-1780 (2004)

Captain Thomas Posey and the 7th Virginia Regiment (2005)

They Are Indeed a Very Useful Corps: American Riflemen in the Revolutionary War (2006)

In This Time of Extreme Danger: Northern Virginia in the American Revolution (2006)

Great Things Are Expected from the Virginians: Virginia in the American Revolution (2008)

To Hazard Our Own Security: Maine's Role in the American Revolution (2010)

Introduction

The man known to history as "Light Horse Harry Lee" (Henry Lee III) was born on January 29, 1756 at Leesylvania, a 2,000 acre Virginia plantation along the Potomac River near the tobacco port town of Dumfries.[1] Young Harry Lee III was born into a distinguished Virginia family. His mother, Lucy Grymes Lee, hailed from Richmond County and was renowned throughout Virginia's tidewater gentry for her attractiveness. Known as the "lowland beauty", folklore said that even a young George Washington was smitten by Lucy's charming features.[2]

Harry Lee's father, Henry Lee II, was a member of one of Virginia's most prominent families. Born in Westmoreland County in 1729, Henry Lee II was a first cousin to such distinguished Virginians as Francis Lightfoot Lee and Richard Henry Lee, both signers of the Declaration of Independence.

Such family connections undoubtedly contributed to Henry Lee II's meteoric rise to political prominence in Prince William County in the 1750's. In 1755, just two years after Lee and his wife settled in Prince William County, he was appointed a Justice of the Peace (local magistrate) and the County Lieutenant (ranking militia officer) by the colonial

---

[1] Ruth and Sam Sparacio, ed., *Virginia County Court Records: Deed and Will Abstracts of Westmoreland County, Virginia: 1747-48*, (McLean, VA: The Antient Press, 1996), 25-26

[2] Edmund Jennings Lee, *Lee of Virginia, 1642-1892: Biographical and Genealogical Sketches of the Descendants of Colonel Richard Lee*, (Baltimore: Genealogical Publishing Co., 1983), 297
Originally published in 1895

governor, Robert Dinwiddie.[3] As County Lieutenant, Lee was responsible for organizing the local militia.

It was probably in this capacity that Henry Lee II first met George Washington, the young commander of Virginia's military forces during much of the French and Indian War. In 1755, Colonel Washington was tasked with defending Virginia's frontier following the crushing defeat of General William Braddock's British army in the woods of Pennsylvania. Washington desperately needed reinforcements and wrote directly to Henry Lee II in October 1755, urging him to send troops to Winchester as quickly as possible.[4] Lee did so and went even further the following spring by accompanying a detachment of county militia to Winchester where he participated in a council of war with Colonel Washington concerning the defense of the frontier.[5]

Unfortunately, further contact between Colonel Washington and Henry Lee II in 1757 and 1758 centered on Washington's disappointment with the number and quality of the Prince William County troops sent to him. On June 19th, 1758, Washington implicitly criticized Henry Lee II when he raised concerns about the Prince William County militia with Virginia's new colonial governor, Francis Fauquier.

---

[3] Ruth and Sam Sparacio, ed., *Virginia County Court Records: Order Book Abstracts of Prince William County, Virginia: 1753-1757* (McLean, VA, 1988), 82, 70

[4] W.W. Abbot, ed., "Colonel George Washington to Henry Lee, 8 October, 1755," *The Papers of George Washington*, Colonial Series, Vol. 2, (Charlottesville: University Press of Virginia, 1983), 87-88

[5] W.W. Abbot, ed., "Council of War, 14-15 May, 1756," *The Papers of George Washington*, Colonial Series, Vol. 3, (Charlottesville: University Press of Virginia, 1984), 129

One hundred Militia...were Order'd from Prince William County but...instead of that number, they sent 73 and every one of them unprovided with either Arms or Ammunition, as the Law directs; by which means they were useless but burthensome to the Country; as they receiv'd true Allowance of Provisions and had their Pay running on. This matter was represented to Colo. Henry Lee, Lieut't of that County, by Sir Jno. St. Clair then Commanding Officer here. The Consequence of this representation was; that about the first of this [month] near 100 Arms were sent up by his order out of which number Scarce 5 were Serviceable; and not more than 30 cou'd possibly by made to Fire. This was also represent'd to Colo. Lee who after professing a Concern for it said, they expect'd Arms from England, (I think) every day, and took no further Acct. of the matter that I have yet heard of. I immediately set Smiths to repairing their Arms, and have at last, with the Assistance of 35 old Muskets which I caus'd to be deliver'd out of the Store here, got this Company, which shou'd consist of 100 Men, (tho' there is but 68) at last completed.[6]

---

[6] W.W. Abbot, ed., "Colonel George Washington to Governor Francis Fauquier, 19 June, 1758," *The Papers of George Washington*, Colonial Series, Vol. 5, (Charlottesville: University Press of Virginia, 1988), 229

Although Washington's criticism of Lee was strong, it does not appear to have had a lasting impact on the relationship between the two men. In future years they developed a friendship that involved many visits and letters and a robust exchange of ideas on one of Washington's favorite topics, agriculture.

By 1758, however, Washington's frustration with the war effort, coupled with his failure to obtain an officer's commission in the British army, led him to resign his command and return to Mount Vernon. He was elected to Virginia's legislative assembly, the House of Burgesses, that same year and was joined by another newly elected member, Henry Lee II.

Lee's representation of Prince William County spanned the next thirty years and included sixteen years in the House of Burgesses (1758-74), participation in all five Virginia Conventions (1774-76), and more than a decade of service in the Virginia state senate (1776-1787). Lee also continued to fulfill his local duties as County Lieutenant, Justice of the Peace, church vestryman, and member of the county committee of correspondence during the Revolution. Such an impressive record of public service could not have gone unnoticed by Lee's eight children.

Although many of Henry Lee II's children followed their father's example and distinguished themselves in public office, Lee's oldest son and namesake, Henry Lee III, surpassed them all. Harry Lee III represented Virginia in the Continental Congress from 1786-88 and forcefully argued in favor of the U.S. Constitution at the state ratifying convention in 1788. He also served as Virginia's governor from 1791-94 and in the U.S. House of Representatives from 1799-1801.

It was on the field of battle during the Revolutionary War, however, that Henry Lee III provided his most significant public service and earned his greatest laurels. The battlefields of New York, New Jersey, Delaware, Pennsylvania, North and South Carolina, and Georgia witnessed Lee's military exploits and heroism, and it was there that Henry Lee III earned the title that he is best known for and one that he would probably be most proud of, "Light Horse" Harry Lee.

Eastern Virginia

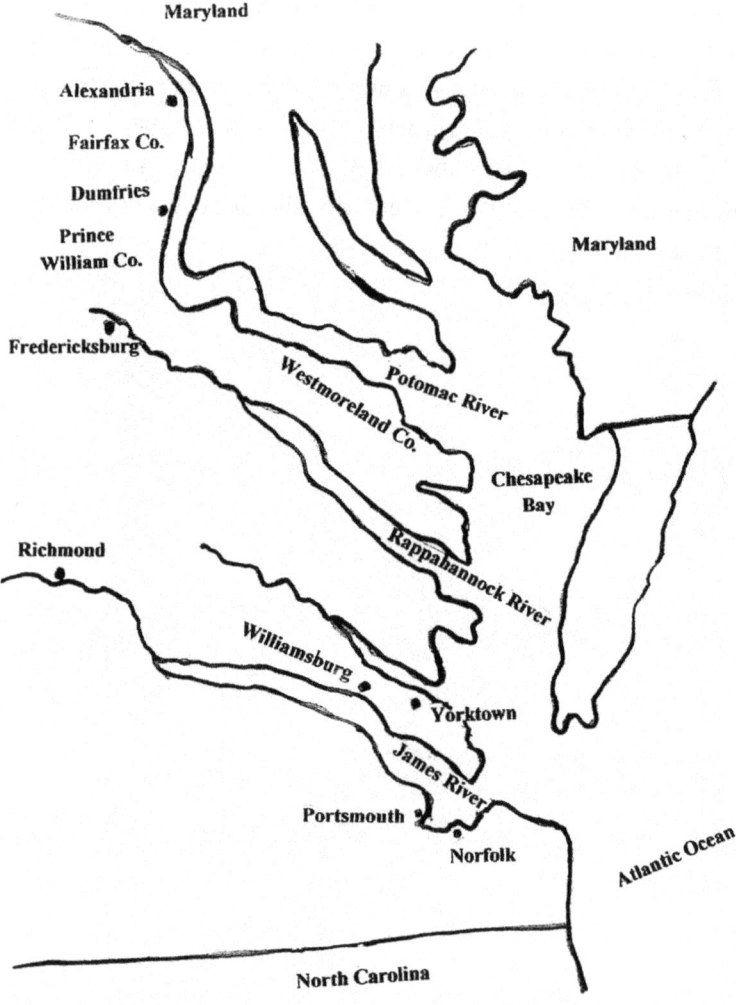

Northern Virginia

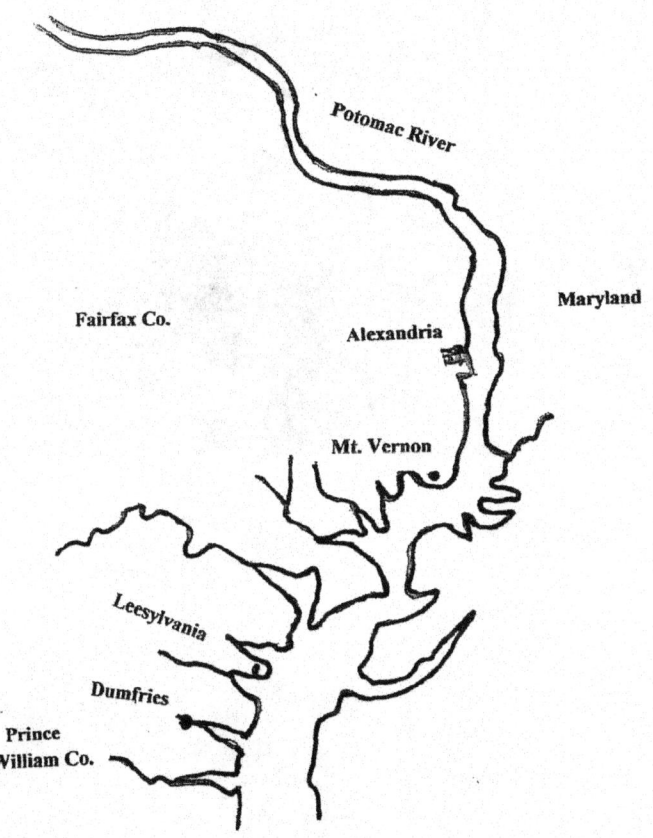

Leesylvania (Artist's Conception)

Courtesy of
Leesylvania State Park

Chapter One

"A great spirit of Liberty actuates every Individual"

Henry Lee III grew up in exciting, yet turbulent times. Raised in the midst of the political turmoil of the 1760's, young Harry witnessed the uproar over the Stamp Act in 1765-66. A few years later, as a teenager, he could proudly point to his father's participation in a non-importation association (boycott) of British goods to protest against the unconstitutional Townshend Duties. Politics, especially opposition to Britain's new colonial policies, was certainly discussed in the Lee household. This was particularly true when the Lee's were visited by Colonel George Washington (something that occurred at least four times between 1768 and 1772).[1] In all likelihood young Harry Lee was impressed by the commanding presence of Colonel Washington, a hero of the French and Indian War.

Unfortunately, no record exists of Colonel Washington's impressions of young Harry Lee. However, others provide a glimpse of the lad. One such person was William Lee, a cousin of Harry's father. In the fall of 1770, Mr. Lee, who had

---

[1] Donald Jackson, ed., *The Diaries of George Washington*, Vol. 2-3, (Charlottesville: University Press of Virginia, 1976 and 1978), Volume 2, 100, 190
Volume 3, 52, 144

moved to England, shared his interest about Harry's education with Harry's father.

> *Your son Harry is a boy of fine parts, and will possess a fine estate, independent of what you may please to give him.... Therefore it surely is incumbent on you to spare no pains or cost to give him a compleat education. This you know cannot be done in Virginia, therefore I wou'd advise his being sent over here [to Britain] immediately for that purpose, and if in so good a work I can be in the least instrumental, I beg you will freely command my services. In this I am the more anxious, as he may one day be the principal representative in Westmoreland [County] of a Family that has shone there for some years, with no small degree of Honor.* "[2]

William Lee was apparently under the impression that young Harry was destined to inherit a large estate in Westmoreland County belonging to Harry's childless uncle, Richard "Squire" Lee. William Lee was concerned about the family's reputation and believed only a quality English education could protect the family's good name. Harry's father disagreed and sent his fifteen year old son to the College of New Jersey in Princeton in 1771.

---

[2] Cazenove Gardner Lee Jr., *Lee Chronicle: Studies of the Early Generations of the Lees of Virginia*, (New York: Vantage Press, 1957), 86-87

Nassau Hall Princeton

By all accounts, Harry Lee excelled at Princeton under the tutelage of Dr. John Witherspoon, the president of the college and an ardent opponent of parliament's recent colonial policies. Lee's classmates included James Madison Jr. and Aaron Burr as well as Harry's younger brother Charles, who joined Harry in his last year of study at Princeton. Harry Lee excelled at his studies and impressed at least one prominent alum of the college. Dr. William Shippen described the young scholar to Richard Henry Lee (Harry's second cousin).

> *Your cousin Harry Lee is in college and will be one of the first fellows in the country. He is more than strict in his morality, has fine genius and is diligent.*[3]

Harry Lee's intellect was displayed for all to see in the fall of 1773 at his commencement ceremony. A description of the ceremony was posted in Rind's *Virginia Gazette*.

> *The 30th of September being the day appointed for the several students in the college of Princeton, in New Jersey, to deliver their discourses, Mr. Henry Lee and Mr. Charles Lee, both of Leesylvania, in Virginia, obtained, the former the degree of batchelor of arts, for his excellent English oration on the liberal arts, and the latter, two of the prizes, for an admirable discourse in the Latin language, and on pronouncing English orations.*[4]

---

[3] Edmund Jennings Lee, *Lee of Virginia*, 329
[4] *Virginia Gazette*, (Purdie) 4 November, 1773

Seventeen year old Harry Lee returned to Leesylvania in the fall of 1773. Tensions between the colonies and parliament had eased during Lee's enrollment at Princeton, due in part to the repeal of the Townshend Duties in 1770, but a dispute over tea was about to explode in Boston. News of the Boston Tea Party in mid-December did not stop Harry and his father from visiting family in Westmoreland County in January 1774. The pair attended a grand ball held by Squire Lee (Harry's uncle) and encountered a former classmate of Harry's from Princeton. Philip Fithian had studied to be a minister while at Princeton, but had recently taken a position as a tutor to the children of Robert Carter of Westmoreland County. Fithian left a detailed description of the ball in his diary.

I was introduced into a small Room where a number of Gentlemen were playing Cards...to lay off my Boots, Riding-Coat, etc. Next I was directed into the Dining-Room to see Young Mr. Lee; He introduced me to his Father – With them I conversed til Dinner, which came in at half after four. The Ladies dined first, when some Good order was preserved; when they rose, each nimblest Fellow dined first – The Dinner was as elegant as could be well expected when so great an Assembly were to be kept for so long a time. –For Drink, there was several sorts of Wine, good Lemon Punch, Toddy, Cyder, Porter etc.—About Seven the Ladies & Gentlemen begun to dance in the Ball-Room – first Minuets one Round; Second Jiggs; third Reels; And last of All Country-Dances; tho' they struck several Marches

> *occasionally – The Music was a French-Horn and two Violins – The Ladies were Dressed Gay, and splendid, & when dancing, their Silks & Brocades rustled and trailed behind them! – But all did not join in the Dance for there were parties in Rooms made up, some at Cards; some drinking for Pleasure; some toasting the Sons of America; some singing "Liberty Songs" as they call'd them, in which six, eight, ten or more would put their Heads near together and roar, & for the most part as unharmonious as an affronted.... I was solicited to dance by several, Captain Chelton, Colonel Lee, Harry Lee, and others,* [but I declined to do so].[5]

Two weeks after the ball, and just days after his 18th birthday, Harry Lee and his uncle, Squire Lee, visited Robert Carter at Nomini Hall. Philip Fithian was present at the dinner and recorded in his diary that, "*The Colonel & Mrs. Carter seem Much pleased with Harry, & with his manner.*"[6]

In spite of all the gaiety among Virginia's gentry class, Virginians, like all of the colonists, were anxious to learn of parliament's reaction to the Boston Tea Party. News of parliament's response reached Virginia in May and created an uproar.

In a series of laws called the Coercive Acts by parliament, but intolerable by the colonists, Britain cracked down hard on Massachusetts. Boston Harbor was closed, the Massachusetts assembly was suspended, and thousands of British troops

---

[5] Hunter Dickinson Farish, ed., *Journal and Letters of Philip Vickers Fithian: A Plantation Tutor of the Old Dominion, 1773-1774*, (Charlottesville: University of Virginia Press, 1999), 56-57
[6] Ibid., 62

under General Thomas Gage occupied Boston to enforce British military rule. Virginians were shocked by the harsh measures and the House of Burgesses expressed its support for Massachusetts by calling for a day of fasting and prayer for the people of Boston. Virginia's royal governor, John Murray, the Earl of Dunmore, deemed the action of the burgesses an insult to the King's authority and dissolved the assembly. Not to be denied, most of the burgesses, including Harry Lee's father, gathered at a Williamsburg tavern and agreed to a non-importation association similar to the one implemented to oppose the Townshend Duties in 1769-70. When calls from the northern colonies for a continental congress reached Williamsburg a few days later, the former burgesses still in the capital gathered and set a date for a special convention to select and instruct Virginia's delegates to the congress.

The 1st Virginia Convention, with Harry Lee's father in attendance, met in Williamsburg in early August 1774 and drafted a boycott proposal to challenge the Intolerable Acts. The convention also selected delegates to represent Virginia in the congress that was to meet in September in Philadelphia. Virginia's Speaker of the House of Burgesses, Peyton Randolph headed a distinguished delegation of Virginians that included George Washington, Patrick Henry, and Richard Henry Lee. The delegates travelled to Philadelphia in early September and hammered out an agreement with the other twelve colonial delegations for a continental wide non-importation agreement. It was hoped that strong economic measures would once again (like the boycott against the Townshend Duties) convince parliament to rescind their unjust colonial policies.

Many colonists, however, believed these economic measures did not go far enough to resist parliament's tyranny. While the 1st Continental Congress met in Philadelphia, county committees throughout Virginia discussed stronger measures of defense. Henry Lee II expressed the view of many Virginians in early October when he asserted that the colonists were prepared to use force if necessary to defend themselves.

> *Unless Britain repeals her late oppressive acts nothing will restore Peace & Harmony but Putting the Colonys on the same footing as they were before the Present Kings Reign.... If Genl Gage Dares to pull a trigger the English troops will rue the Day, the Colonys are determined as One Man if things should be reduced to Extremitys, to repell force by force....*[7]

Just how Virginia would exercise force remained to be seen. Lord Dunmore's dissolution of the House of Burgesses in late May had prevented the Virginia assembly from renewing the colonial militia law of 1771. This meant that authority to muster the militia in Virginia no longer existed. With thousands of British troops in Boston and tension between the colonies and parliament extremely high, more than a few Virginians believed that stronger measures of defense were necessary. Fairfax County, at the behest of George Mason, led the way by forming an extra-legal independent militia company of gentlemen volunteers who

---

[7] "Henry Lee II to William Lee, 1 October, 1774," Lee-Ludwell Papers, Virginia Historical Society

pledged to supply themselves with the necessary equipment of a soldier and muster regularly to master the art of war.[8]

Prince William County followed Fairfax County's lead in November and formed their own independent militia company. Both companies spent the winter and spring of 1774-75 acquiring arms and accoutrements and practicing the military drills of the day.

Interestingly, although his father was the County Lieutenant of Prince William County and many of his neighbors were members of the independent company, eighteen year old Harry Lee III was apparently not a member of the Prince William County independent militia company. Instead, young Harry Lee was preparing to study law in England.[9] Dramatic events in the spring of 1775, however, altered these plans.

Virginia, like all of the colonies in America, was on edge in the spring of 1775. Talk of armed conflict with Britain spurred many to prepare for war. Harry Lee's father noted such preparations in a letter to his cousin, William Lee, in England.

---

[8] Robert A. Rutland, ed., "Fairfax County Militia Association, 21 September, 1774," *The Papers of George Mason*, Vol. 1, (University of North Carolina Press, 1970), 210-11

[9] Henry Lee, *The Revolutionary War Memoirs of General Henry Lee*, (NY: Da Capo Press, 1998), 16
Originally published in 1812 as, *Memoirs of the war in the Southern Department of the United States* and republished in 1869 as *The Revolutionary War Memoirs of General Henry Lee*, edited by Robert E. Lee. The 1998 edition is an unabridged republication of the third edition published in 1869.

> *We are making large Quantitys of Salt Petre from the Nitre in the Tobacco Putrified with Urine and have made some very strong well grained Powder in this County...which ketches quick and shoots with great force, so that we shall be able in Future to supply ourselves with Salt Petre and gun Powder without Importing any.... The Gentlemen are training themselves thro' the Continent every week and have raised Companys who muster two days every week and emulate to Excell each other in the manual maneuvers and Evolutions as practiced by the King of Prussia's Troops, for we are determined on Preserving our Libertys if necessary at the Expense of our Blood, being resolved not to survive Slavery. You may rely on it that the Continental Association will be most sacredly Kept as the County Committee will not suffer the least breach to pass unnoticed and are very watchful.*[10]

While militia companies drilled throughout Virginia and the boycott was strictly enforced, delegates to the 2nd Virginia Convention, including Henry Lee II, met at St. John's Church in Richmond in mid-March. The most important issue of the meeting was introduced on March 23rd when Patrick Henry proposed that Virginia *"Be immediately put into a posture of defense."*[11] Henry believed that armed conflict with Britain was inevitable, and he called on Virginia to assume a war footing.

---

[10] Edmund Jennings Lee, "Henry Lee II to William Lee, 1 March, 1775," *Lee of Virginia*, 293-94

[11] William J. Van Schreeven and Robert L. Scribner, eds., *Revolutionary Virginia: The Road to Independence,* Vol. 2 (University Press of Virginia, 1975), 366-367

Henry's proposal was opposed by moderates like Edmund Pendleton and Robert Carter Nicholas. They argued that it was too confrontational and would provoke Parliament. Patrick Henry rose to address his critics and, in doing so, delivered one of the greatest speeches in American history. Henry was alarmed by Britain's military buildup in Massachusetts and asked the delegates a pointed question:

> *Are fleets and armies necessary...[for] reconciliation? Have we shown ourselves so unwilling to be reconciled, that force must be called in to win back our love? Let us not deceive ourselves, sir. These are the implements of war and subjugation—the last arguments to which kings resort...I ask gentlemen, sir, what means this martial array, if its purpose be not to force us to submission...Has Great Britain any enemy in this quarter of the world, to call for all this accumulation of navies and armies? No, sir, she has none. They are meant for us, they can be meant for no other. They are sent over to bind and rivet upon us those chains which the British ministry have been so long forging....*[12]

Henry recounted the failure of the numerous pleas and petitions sent to Britain over the past decade and confidently exclaimed that, *"Three million people, armed in the holy cause of liberty...are invincible by any force which our enemy can send against us."*[13] He then asserted that conflict was inevitable and urged the delegates to prepare the colony for it:

---

[12] William Wirt, *The Life of Patrick Henry*, (New York: M'Elrath & Bangs, 1832), 139
[13] Ibid., 141

There is no retreat, but into submission and slavery! Our chains are forged. Their clanking may be heard on the plains of Boston! The war is inevitable – and let it come!! I repeat it sir, let it come!! It is vain, sir, to extenuate the matter. Gentlemen may cry, peace, peace – but there is no peace. The war is actually begun! The next gale that sweeps from the north will bring to our ears the clash of resounding arms! Our brethren are already in the field! Why stand we here idle? What is it that gentlemen wish? What would they have? Is life so dear, or peace so sweet, as to be purchased at the price of chains and slavery? Forbid it, Almighty God! I know not what course others may take; but as for me, give me liberty, or give me death![14]

Patrick Henry's stirring appeal worked, and his resolution passed by a narrow margin.

The mood of many Virginians in the spring of 1775 was defiant and Henry Lee II noted a week after Henry's speech that, "*a great spirit of Liberty actuates every Individual.*[15] This "spirit of liberty" was evident at Mount Vernon when Charles Lee paid a visit to George Washington on April 16th, 1775. Lee, who was no relation to Harry Lee or the Lee's of Virginia, had only recently moved to the colonies from England in 1773. Although he was very eccentric in his manners, he had a fine reputation as an experienced and knowledgeable former British officer (retired). Lee travelled extensively throughout the colonies in 1773-74 and was admired by many colonists, including Colonel George Washington, who met with Lee in Philadelphia at the First Continental Congress and quite possibly served with Lee in the French and Indian War. Lee visited Mount Vernon in

---

[14] Ibid.
[15] Edmund Jennings Lee, "Henry Lee II to William Lee, 1 April, 1775," *Lee of Virginia*, 292

December 1774 and again a few months later in April and the two men seemed to have developed a comfortable relationship. Lee genuinely embraced the views and opposition of many colonists to Britain's colonial policies, and with conflict seemingly imminent, it was reassuring to many Virginians to have the support of such an experienced military officer.

Lee and Washington were joined at Mount Vernon by nineteen year old Harry Lee III who, by chance, was visiting Colonel Washington when Charles Lee arrived. It is difficult to characterize with any certainty what occurred or was discussed at this gathering in mid-April as no one recorded their impressions, but one would assume that young Harry Lee was awestruck to be in the presence of such distinguished men.

Harry Lee's high regard for Charles Lee was evident in a letter the young man wrote three months later to Lee on July 5th, 1775. Much had occurred in the colonies since Harry Lee's visit to Mount Vernon. Blood had been shed in Massachusetts (Lexington and Concord and Bunker Hill) and Governor Dunmore had fled from Williamsburg with his family for the safety of a British naval ship. The formation of the continental army and the appointment of General George Washington to command it were also significant developments as was the appointment of General Charles Lee to the American army.

It was the appointment of Charles Lee to the continental army that prompted young Harry Lee to write to him. After expressing his pleasure at Lee's appointment, Harry Lee made a request.

> *The familiarity with which you treated me in Virginia has induced me to ask* [permission] *to enlist under your command in order to acquaint myself with the art of war.*[16]

No longer interested in the study of law in England, nineteen year old Harry Lee wanted to join General Lee and the newly created continental army at Boston and serve as Lee's aide de-camp. Harry Lee's background, education, and temperament seemingly made him a fine choice for such a position but he was disappointed in his request because General Lee never received the letter. It was intercepted by the British.

While young Harry impatiently waited at Leesylvania for a reply to his request that was not forthcoming, the 3rd Virginia Convention met to strengthen Virginia's military forces against Lord Dunmore and the British. The Convention had assumed the reins of government after Governor Dunmore's flight in June and by August 1775 had restructured the colony's military.

Concerned about the reliability, effectiveness, and discipline of the county militias, the Convention decided to form two regiments of regular (full time) soldiers to serve for one year.[17] The last time this was done was during the French and Indian War, twenty years earlier. The Convention also bolstered the colony's defenses with sixteen battalions of minutemen. These men were drawn from the ranks of the militia and were *"more strictly trained to proper discipline"* than the ordinary militia.[18]

---

[16] "Henry Lee III to Charles Lee, 5 July, 1775," Thomas Gates Papers, William L. Clements Library, Ann Arbor, Michigan.
[17] William W. Hening, ed., *The Statutes at Large Being a Collection of all The Laws of Virginia*, Vol. 9, (Richmond: J & G Cochran, 1821), 10, 16
[18] Ibid., 16

The 3rd Virginia Convention completed its military preparations by establishing a new militia law that decreed that,

> *All male persons, hired servants, and apprentices, above the age of sixteen, and under fifty years...shall be enlisted into the militia...and formed into companies....*[19]

Unfortunately, little evidence exists to determine what military service Harry Lee III performed in the months leading up to his appointment as a dragoon officer in June 1776. It is likely that by the fall of 1775 Lee, disappointed by the lack of a reply from General Lee, would have at the very least mustered with the local militia in accordance to the 3rd Convention's decree. And given the stature of his family and the fact that he was appointed captain of a troop of dragoons in June 1776, it is likely that Lee served as a company grade officer, (ie. ensign, lieutenant, or even captain) while in the militia. If so, he was probably terribly disappointed with the experience because there was little in the way of weapons and supplies left for the militia after the regular troops and minute-men received their allotment. Lund Washington, a relative of General Washington and the caretaker of Mount Vernon while Washington was away, noted the absence of weapons among the local militia in neighboring Fairfax County in early 1776.

> *The Minute Battalion is gone to Williamsburg & with them almost all the guns that were worth haveg in the County. Our Militia Exercises with Clubs, if they come to close quarters in an engagement they perhaps may do some Execution but not otherwise....*[20]

---

[19] Ibid., 27-28
[20] Philander D. Chase, ed., "Lund Washington to George Washington, 17 January, 1776," *The Papers of George Washington*, Vol. 3, 129-130

If Harry Lee served in the Prince William County militia (as was required by law) he would have probably faced the same challenges and frustrations.

Relief from the disappointment of militia service arrived for Harry Lee in the form of General Charles Lee, who was ordered to Virginia in March 1776. General Lee, who had spent most of 1775 in Massachusetts with General Washington and the American army, was tasked with improving Virginia's defensive preparations. Young Harry Lee was thrilled to learn of General Lee's new command and seized the opportunity in early April to once again offer his service to the general.

> *I must assure you, that I shall be particularly attentive to the discharge of any office you may think proper to honour me with.*[21]

It is difficult to determine why Harry Lee did not join General Lee's military family as an aide-de-camp, but one possible reason is that he was swayed by General Lee's public appeal in late April for Virginia's young gentlemen to form companies of light dragoons.[22] One can easily envision twenty year old Harry Lee embrace such an appeal, he was, after all, the epitome of a young Virginia gentleman.

In mid-May, the 5th Virginia Convention acted on General Lee's proposal and approved the following ordinance.

> *That four Troops of Horse be raised for the better Security and Defence of this Colony that the Officers and Troopers at their own Expence provide their horses, Arms and Accoutrements and be allowed a reasonable pay and proper Subsistance and be paid*

---

[21] -------- "Henry Lee to Charles Lee, 7 April, 1776," *Collections of the New York Historical Society*, Vol. 4, 1871, 391
[22] *Virginia Gazette*,(Purdie) 26 April and 3 May, 1776

for such Horses as are killed or taken by the Enemy.[23]

Three weeks later, on the same day that Richard Henry Lee introduced Virginia's resolution for independence in the Continental Congress in Philadelphia, the 5th Virginia convention increased the sections or troops of cavalry from four to six and selected officers for each troop.[24] Thirty-four year old Theodorick Bland was chosen to command the 1st troop of cavalry and was therefore, the overall commander of the six troops of dragoons.[25] Although he had little military experience, Bland's family connections helped overcome whatever concerns the delegates may have had. The captains for the next three troops were all former captains of minute companies. Twenty year old Henry Lee III was considered for command of the 3rd and 4th troop, but finished third and second in the vote, respectively.[26] Lee was finally selected to command the 5th troop. Interestingly, future Chief Justice of the Supreme Court John Marshall was almost elected lieutenant of the 5th troop, but lost out to John Belfried.[27] Henry Peyton of Loudoun County, who would fight under Lee for most of the war, was chosen to be the cornet (ensign) of the 5th troop.[28]

---

[23] Tarter and Scribner, eds., "Proceedings of the 5th Virginia Convention, 20 May, 1776," *Revolutionary Virginia, The Road to Independence*, Vol. 7 Part 1, 194

[24] Tarter and Scribner, eds., "Proceedings of the 5th Virginia Convention, 7 June, 1776," *Revolutionary Virginia, The Road to Independence*, Vol. 7 Part 1, 390-395

[25] Tarter and Scribner, eds., "Proceedings of the 5th Virginia Convention, 13 June, 1776," *Revolutionary Virginia, The Road to Independence*, Vol. 7 Part 2, 474-75

[26] Ibid., 475

[27] Ibid., 477

[28] Ibid., 489

Captain Lee's cavalry troop, like the other five, consisted of 30 rank and file privates, 3 corporals, and a trumpeter.[29] The Convention originally expected each cavalryman to bare the cost of their own horse, arms, and accoutrements, but this policy quickly changed when few troopers volunteered under such terms. To encourage enlistments the convention announced in late June that,

> *Each trooper shall be furnished with the following arms and accoutrements...a carbine with bucket and straps, a pair of horseman's pistols and holsters, a tomahawk, a spear, and a good saddle.... And be it further ordained, that instead of the corporals, trumpeters, and private troopers, furnishing their own horses, arms, and accoutrements...the said horses, arms, and accoutrements, shall be furnished at the expense of the publick.*[30]

To partially compensate for the significant expense of this new policy, the convention reduced the pay of the troopers and extended their enlistments by over a year to December 1, 1778.[31]

The Convention expected Lee and his fellow officers to recruit and equip the necessary men as quickly as possible. No record exists of Captain Lee's recruitment efforts, but the recruits of Lee's lieutenant, John Belfield, were reviewed and accepted into the service in late August.[32]

Two weeks earlier, Virginians rejoiced at the news that Lord Dunmore and his force of runaway slaves, Tories, and Redcoats had sailed out of Chesapeake Bay. Dunmore's departure allowed Virginia's leaders to finally send troops

---

[29] *Virginia* Gazette, (Purdie) 14 June, 1776, 4
[30] *Virginia* Gazette,(Purdie) 28 June, 1776, 1
[31] Ibid.
[32] H. R. McIlwaine, ed., "31 August, 1776," *Journals of the Council of the State of Virginia*, Vol. 1, 144

north to reinforce General Washington's army in New York. Thousands marched out of the state in the fall, but the Virginia dragoons did not go with them. It appears that Captain Lee and the rest of the Virginia cavalry spent the remainder of 1776 in the vicinity of Williamsburg. Construction of a barracks and stables in Williamsburg to accommodate one hundred horses (nearly half of the Virginia cavalry) was authorized by the government in September.[33] Orders were also given in the fall to the keeper of the powder magazine in Williamsburg to supply Captain Lee's troop with cartouch boxes to hold their ammunition.[34]

While Captain Lee and his fellow cavalrymen equipped and trained themselves in Williamsburg, General Washington and his army struggled to defend American independence and hold back a vastly stronger British army in New York and New Jersey. By the time the Virginia light cavalry was ordered north to join Washington's army in mid-December, the situation for General Washington and his men appeared grim.[35]

---

[33] Ibid., "26 September, 1776," 175
[34] Ibid., "15 October, 1776," 198
    "9 November, 1776," 233
[35] Ibid., "18 December, 1776," 288

Virginia Light Dragoon

Courtesy of
Pamela Patrick White

Chapter Two

"He is so enterprising and useful an officer"

Governor Patrick Henry's order in December for the Virginia light cavalry to march north and join General Washington's army sparked a flurry of activity in preparation for the 300 mile march to Philadelphia. While Captain Lee and his fellow cavalrymen prepared for the march, General Washington and his demoralized army struggled to protect the de facto American capital.

The war had gone poorly for the Americans since their declaration of independence in July and many worried that Philadelphia would fall to British General Lord Howe just as New York had in September. Even worse, morale among the Americans was so low in the last month of 1776 that many doubted there would even be an American army left to face the British in 1777. With the enlistments of half his tiny army due to expire on January 1st, and the local militia reluctant to turn out, General Washington was desperate for reinforcements. He candidly described the grim situation to his brother on December 18th.

> *I have no doubt but that General Howe will still make an attempt upon Philadelphia this Winter – I see nothing to oppose him in a fortnight from this time, as the term of all the Troops except those of Virginia, (reduced almost to nothing) and Smallwoods Regiment from Maryland (in the same condition) will expire in that time.*

> *In a word my dear Sir, if every nerve is not strain'd to recruit the New Army with all possible expedition, I think the game is pretty near up....*[1]

Fortune finally shined upon General Washington and his army a week later when they scored a stunning and decisive victory over a large Hessian force in Trenton, New Jersey. Washington's bold crossing of the Delaware River on Christmas night and his successful surprise attack upon Trenton the following day boosted American morale. General Washington and his men followed this crucial victory with another eight days later at Princeton.

By the time Captain Lee and the Virginia light cavalry reached Philadelphia in mid-January, the situation in New Jersey had improved considerably for the Americans. Washington's decisive actions at Trenton and Princeton prompted the British to pull back from western and central New Jersey. They still held a chain of posts in the eastern part of the state, but their reach across New Jersey had diminished significantly.

Although the size of the American army in the winter months of 1777 was alarmingly small, General Washington kept pressure on the British by harassing their foraging parties whenever he could. Detachments were posted along the Watchtung Mountains to screen the main American encampment at Morristown and strike at enemy patrols and forage parties whenever the opportunity presented itself. When the Virginia cavalry reached Philadelphia in January they were briefly assigned to a detachment under the command of General William Alexander (Lord Stirling) of New Jersey. General Stirling was posted at Baskenridge and was under orders to,

---

[1] Philander D. Chase, ed., "General Washington to Samuel Washington, 18 December, 1776," *The Papers of George Washington*, Vol. 7, 370

> *Harass and annoy the Enemy by keeping Scouting parties constantly (or as frequently as possible) around their Quarters.*[2]

The Virginia light cavalry was ideally suited for such activities and in all probability served in many of these scouting parties. They also participated in a few larger operations. In early February, Major Bland, who had been promoted by the Virginia government back in June, led a detachment of cavalry in a large foraging expedition to, "*remove out of* [the enemy's] *reach all of the Horses, Waggons & fat Cattle...lying between Quibble Town & the Sound, eastward; approaching as near the Enemy as you can in safety.*"[3] Major Bland described the outcome of the expedition in a letter to his wife.

> *I was two days ago with part of my regiment, and a body of troops under the command of General Sullivan, on a foraging party...one or two of the light horse fired a shot or two at a small party of the enemy; a party of foot marched up to attack them, but they retreated and left us the field without the least damage done to either side. We brought off five or six hundred cattle, and about as many sheep, belonging to tories.*[4]

---

[2] Frank E. Grizzard Jr., ed., "Orders to Major General Stirling, 4 February, 1777," *The Papers of George Washington*, Vol. 8, 245

[3] Grizzard Jr., ed., "General Washington to General Sullivan, 3 February, 1777," *The Papers of George Washington*, Vol. 8, 237

[4] Charles Campbell, ed., " Theodorick Bland Jr. to his lady," *The Bland Papers: Being a Selection from the Manuscripts of Colonel Theodorick Bland Jr. of Prince George County, Virginia*, (Petersburg: Edmund & Julian Ruffin, 1840), 47

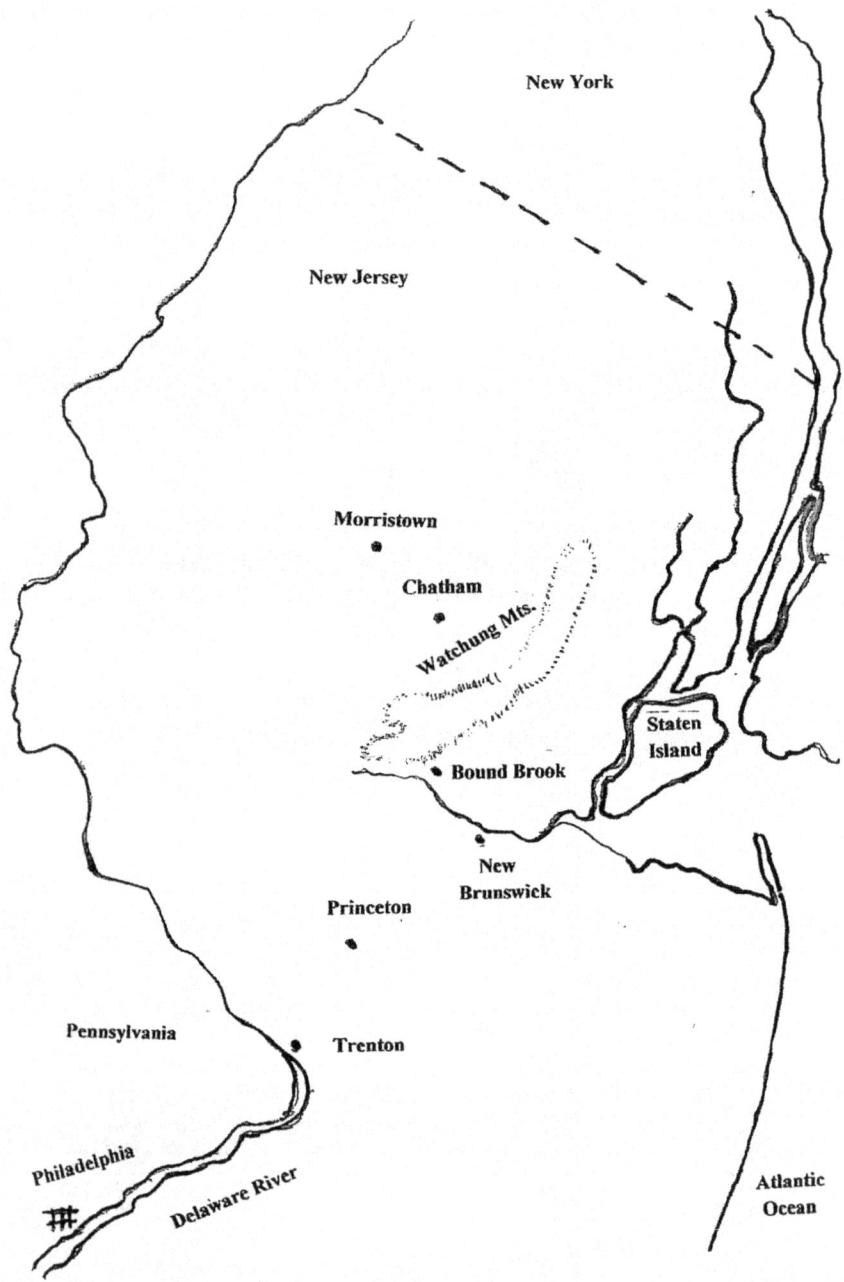

The small size and weak condition of Washington's army in early 1777 limited his ability to conduct such large operations on a regular basis. As a result, Washington's troops, particularly the Virginia cavalry, spent most of their time on small patrols, frequently skirmishing with enemy patrols and forage parties. Captain Johann Ewald of the German Jagers (riflemen) remarked in his diary that in a six week period from mid- February through March,

> *Nothing important happened due to the constant high snow, except for the daily skirmishing of our patrols and the continual alarms of the outposts on both sides. Scarcely a day passed when we did not have to stand under arms for hours in the deepest snow.*[5]

General Washington made a similar observation in April.

> *We have greatly harassed, & distressed the Enemy, by continually skirmishing with their Foraging Parties, and attacking their Picquet Guards.*[6]

The constant pressure that General Washington maintained on the enemy was quite an accomplishment for his vastly outnumbered army and something that Bland's Virginia light horse contributed significantly to. With three other continental cavalry regiments not yet complete or in camp, the workload on Major Bland's dragoons became exhausting. They were posted at several locations in New Jersey and were constantly on duty. General Washington acknowledged the toll the constant patrols took on Bland's cavalry in a letter to General Alexander McDougall of New York in April. McDougall had asked Washington to divert some of the continental cavalry being raised in Connecticut to New York

---

[5] Johann Ewald, *Diary of the American War: A Hessian Journal*, (New Haven and London: Yale University Press, 1979), 55

[6] Philander D. Chase, ed., "General Washington to Landon Carter, 15 April, 1777," *The Papers of George Washington*, Vol. 9, 171

to help him defend the New York Highlands. Washington's reply was firm and revealing.

> *I cannot at this time spare any of the Continental Light Horse raising in Connecticut, they are much wanted here, those we have* [Bland's] *having* [been] *greatly reduced by the constant service since they joined me.*[7]

While many of Major Bland's cavalrymen were posted at Morristown and Chatham, Captain Lee and his cavalry troop spent much of the winter and early spring attached to General Benjamin Lincoln's force at Bound Brook, just seven miles from enemy held New Brunswick. The high New Jersey hills provided an excellent perch to observe the enemy and Lee's close proximity to the British led to frequent skirmishes.

In late March, word arrived that Congress had re-designated the Virginia light cavalry the 1st Continental Light Dragoon Regiment.[8] Congress had placed the Virginia light dragoons onto the continental establishment two months earlier in mid-January, so the new unit designation was not a significant change.[9] It mostly meant promotions for a few of the officers. Theodorick Bland was promoted to colonel of the new regiment while Benjamin Temple and John Jameson were promoted to lieutenant colonel and major, respectively.[10] Captain Lee did not receive a promotion, but his cornet, Henry Peyton, was promoted to lieutenant of Lee's troop in the place

---

[7] Chase, ed., "General Washington to General McDougall, 17 April, 1777," *The Papers of George Washington*, Vol. 9, 187
[8] Chase, ed., "General Orders, 31 March, 1777, Footnote 3," *The Papers of George Washington*, Vol. 9, 25
[9] Worthington C. Ford, ed., "14 January, 1777," *Journals of the Continental Congress*, Vol. 7, 34
[10] Chase, ed., "General Orders, 31 March, 1777," *The Papers of George Washington*, Vol. 9, 25

of John Belfield who had become captain of his own troop of cavalry.[11]

The advent of warm weather in April sparked increased activity in the two armies and more action for Captain Lee and his troop, all of which took a toll on Lee's men and horses. When Captain Lee received orders in late April to march to Chatham via Morristown, the image conscious young officer expressed concern to Colonel Bland about the ragged appearance of his tired men.

> *How happy would I be, if it was possible for my men to be furnished with caps and boots, prior to my appearance at head quarters. You know, dear colonel, that, justly, an officer's reputation depends not only on the discipline, but appearance of his men. Could the articles mentioned be allowed my troop, their entrance into Morris would secure me from the imputation of carelessness, as their captain, and I have vanity to hope would assist in procuring some little credit to their colonel and regiment.*[12]

Lee's comments were actually well founded, for just eleven days earlier General Washington had sent a circular letter to a number of Virginia commanders, including Colonel Bland, complaining about the condition and appearance of their men.

> *I am informed, and indeed I have observed, that the men of your Regiment are so exceedingly bare of necessaries that it not only contributes to their unhealthiness, but renders them absolutely unfit to take the field. Inattention to the Wants of Soldiers marks the bad officer – it does more, it reasonably removes that Confidence on which the officer's*

---

[11] Ibid.
[12] Campbell, "Henry Lee to Col. Bland, 18 April, 1777," *The Bland Papers,* 51

Honour & Reputation must depend – As there is Cloathing now here, I desire you may immediately cause inquiry to be made into what is wanting, and make returns, that if the things wanted are not here, they may be ordered on.[13]

As Captain Lee passed through the main American encampment at Morristown in late April he likely noticed a revitalized American army whose ranks swelled with new recruits. Encouraged by these reinforcements, General Washington held a council of war in early May to solicit the advice of his generals. They counseled against any aggressive offensive action and supported the continuation of a defensive posture.[14]

This preference for the status quo did not extend to Washington's organization of the army, which was formed into ten new brigades. Each brigade, which ranged between 700 to 1,000 men, consisted of four or five undermanned regiments, typically from the same state. By mid-May the combined strength of these brigades surpassed 8,000 strong.[15]

General Washington's continental cavalry was raised on a similar regional basis. Congress had authorized the formation of four continental cavalry regiments during the winter of 1776-77, but only one, Colonel Bland's 1st Continental Dragoons, had seen significant service in Washington's army. Colonel Stephen Moylan's 4th Continental Dragoons (raised in the mid-Atlantic region and Virginia) and Colonel George Baylor's 3rd Continental Dragoons (raised in Virginia and

---

[13] Chase, ed., "Circular to the Commanding Officers of Several Virginia Regiments, 7 April, 1777," *The Papers of George Washington*, Vol. 9, 83

[14] Chase, ed., "Council of War, 2 May, 1777," *The Papers of George Washington*, Vol. 9, 324

[15] Chase, ed., "Arrangement and Present Strength of the Army in New Jersey, 20 May, 1777," *The Papers of George Washington*, Vol. 9, 492-93

Maryland) each had sent a troop of cavalry to Washington by June, but most of the dragoons in these regiments, along with Colonel Elisha Sheldon's entire 2nd Continental Dragoon Regiment (raised in Connecticut, Massachusetts and New Jersey) had yet to arrive in camp.[16]

The continued absence of the bulk of the continental cavalry placed an enormous burden on Colonel Bland's horsemen. Scattered among numerous posts in New Jersey and constantly on patrol, the endless duty wore down the Virginia cavalrymen and their horses to such a point that General Washington agreed to temporarily withdraw Bland's regiment from service. Washington discussed his decision in a letter to General John Sullivan.

> *The Virginia Regiment of Horse had been so detached the whole Winter that I could not deny Colo. Bland his request to draw them together that they may be properly equipped, which they have never yet been.*[17]

To replace Bland's exhausted dragoons, Washington ordered Colonel Elisha Sheldon, who was still organizing his cavalry regiment in Connecticut, to hurry his men along.

> *The Virginia regt of Light Horse have been so worn down by hard service that* [unless] *they are reliev'd of part of their Duty they will be totally unfit for service of any kind. I therefore desire that you will*

---

[16] Chase, ed. "General Washington to Colonel Baylor, 17 May, 1777," and "General Washington to Colonel Sheldon, 24 May, 1777," *The Papers of George Washington*, Vol. 9, 448 and 521 and "General Washington to General Schuyler, 16 June, 1777," *The Papers of George Washington*, Vol. 10, 54

[17] Chase, ed., "General Washington to General Sullivan, 7 June, 1777," *The Papers of George Washington*, Vol. 9, 639

> send on ev'ry man of your Regt that is cloathed & Mounted and that have had the small Pox.[18]

Colonel Bland's exhausted cavalry regiment was temporarily removed from active service in July to rest and recover from the hard winter and spring.[19] A top priority was to allow the regiment to, *"get forage to recruit their horses."*[20] Unfortunately for the Virginia dragoons and their mounts, their respite ended abruptly when General Washington ordered the army northward to New York in mid-July in anticipation of a British move in that direction.

Reports that the bulk of General Howe's army had boarded transport ships in New York harbor convinced Washington that Howe intended to sail up the Hudson River and cooperate in a joint operation with a 7,000 man British army from Canada under General John Burgoyne. A juncture between these two British forces, combined with the British navy's dominance of the ocean, would effectively sever New England from the rest of the states. General Washington moved to counter this threat by marching north into New York.

While Washington and the bulk of the American army marched to New York, Colonel Bland received orders to post his cavalry regiment at Bound Brook, New Jersey.

> *His excellency...thinks it not improbable the enemy may take it into their heads to make some incursion into the Jerseys, to plunder and distress the inhabitants; or, perhaps, even to endeavor to destroy our stores at Morris-Town. This will be worthy of your attention; and should it happen, you will give all the assistance in your power to the force which may*

---

[18] Chase, ed., "General Washington to Colonel Sheldon, 9 June, 1777," *The Papers of George Washington*, Vol. 9, 652-53

[19] Frank E. Grizzard Jr.,, "General Orders, 1 July, 1777," *The Papers of George Washington*, Vol. 10, 162

[20] Ibid.

> be collected to oppose them. Use every expedient you can think of to gain the exactest intelligence possible of the movements of the enemy. Our situation and theirs are such, that it is extremely difficult to know what they are about, and we are rather in the dark with respect to it.[21]

Washington's uncertainty was prompted by the disappearance of General Howe and the British fleet. Instead of sailing up the Hudson River, Howe sailed out to sea, making it impossible for Washington to determine his true destination. After days of uncertainty, Washington received news of Howe's location in a dispatch from Congress.

> I receiv'd Information from Congress that the Enemy were actually at the Capes of Delaware – This brought us in great haste [southward] for the defence of [Philadelphia] but in less than 24 hours after our arrival we got Accts of the disappearance of the Fleet [again].[22]

Amazingly, the British fleet had sailed back out to sea and disappeared over the horizon again. General Washington was extremely perplexed and waited with the army near Philadelphia for Howe to reappear.

> We remain here in a very irksome state of Suspence, Some imagining that they are gone to the Southward, whilst a Majority (in whose opinion...I concur) are satisfied they are gone to the Eastward [towards New England]. The fatigue, however, & Injury, which Men must Sustain by long Marches in such extreme

---

[21] Campbell, "From Col. Alexander Hamilton, 21 July, 1777," *The Bland Papers*, 61

[22] Grizzard, Jr., ed., "General Washington to John Augustine Washington, 5-9, August, 1777," *The Papers of George Washington*, Vol. 10, 514-515

> *heat as we have felt for the last five days, must keep us quiet till we hear something of the destination of the Enemy.*[23]

Three more weeks passed before General Howe's plans were finally revealed. Washington received an urgent dispatch on August 22nd reporting that Howe's fleet was spotted high up into the Chesapeake Bay.[24] General Howe's intentions were finally clear, he meant to attack Philadelphia from the southwest. General Washington immediately ordered his troops to march south, through Philadelphia, to face the enemy.

On the day the army departed, Captain Henry Lee experienced an embarrassing ordeal, a court martial. It is unclear who charged Captain Lee with disobedience of orders or what the circumstances of the incident were, but the verdict of the court, which was delivered two days later on August 25th, could not be clearer.

> *The Court having fully considered the charge and evidence are unanimously of opinion, that Capt. Henry Lee is not guilty of the charge exhibited against him, and do acquit him with honor – And they are also unanimously of opinion, that the charge against Capt. Lee is groundless and vexatious, and that Capt. Lee, in what he did, acted merely for the good of his troop – The Commander in Chief approves of Capt. Lee's acquittal with honor.*[25]

Captain Lee spent little time celebrating the verdict of the court; he was eager to return to his command and rushed

---

[23] Ibid.
[24] Philander D. Chase and Edward G. Lengel, eds., "General Washington to General Sullivan, 22 August, 1777," *The Papers of George Washington*, Vol. 11, 48
[25] Chase and Lengel, eds., "General Orders, 25 August, 1777," *The Papers of George Washington*, Vol. 11, 63

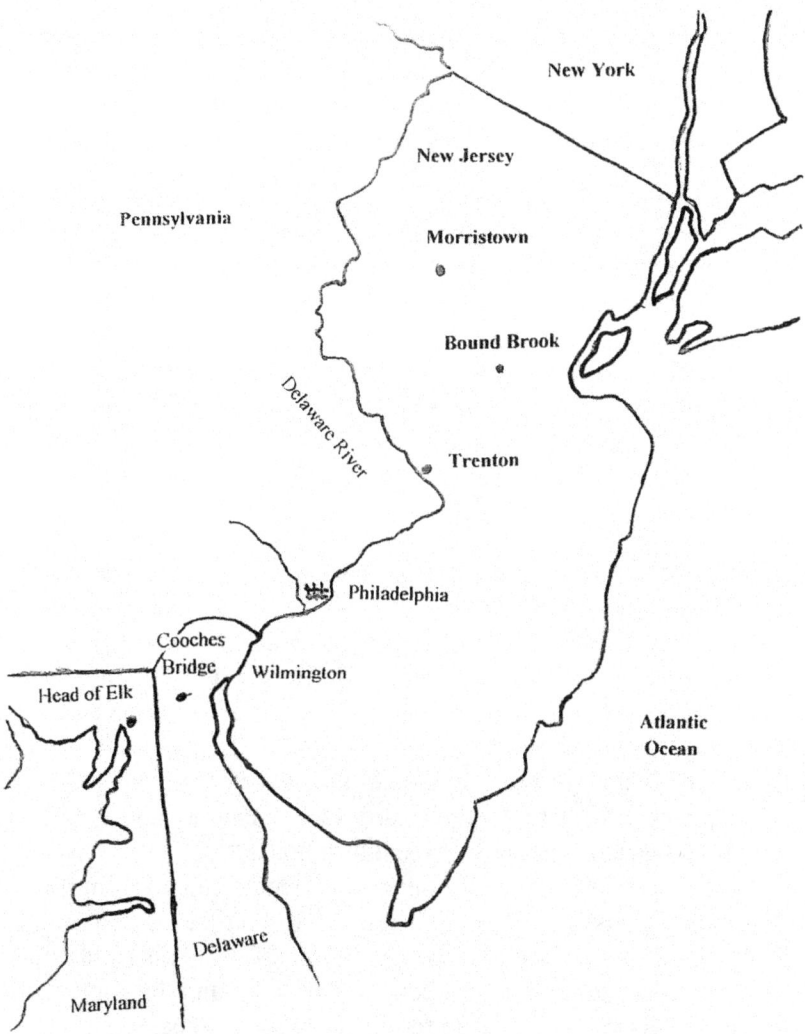

southward to rejoin his cavalry troop. With the British army only 50 miles from Philadelphia, General Washington needed to redeploy his army, but he was uncertain of the exact route Howe planned to take to Philadelphia. To help determine this, Washington instructed Colonel Bland to post his cavalrymen as close to the enemy as possible to observe their movements.

> *Such is the importance of our having early notice of their beginning to move, that I must repeat to you the necessity there is of keeping small guards and constant patrols, both of horse and foot, on the flanks and in front of the enemy, as near to them as prudence will permit, so that they cannot possibly move any way, without your having information of it. I shall expect to have immediate notice of every matter of importance which comes to your knowledge....*[26]

Apparently not content to wait for reports from Colonel Bland and the other scouting parties, General Washington, escorted by a detachment of dragoons, rode towards Head of Elk, Maryland to observe the enemy for himself. With him were General Nathanael Greene of Rhode Island and recently arrived French aristocrat and volunteer the Marquis de LaFayette. The young Frenchman recalled years later that a violent storm forced the small party to risk capture and spend the night near the enemy in a local farmhouse.[27]

While General Washington's impatience caused him to imprudently risk capture, his cavalry officers executed his order to patrol as close to the enemy as possible. This resulted in numerous encounters between the two sides and the capture of enemy troops who strayed too far from their units.

---

[26] Chase and Lengel, eds., "General Washington to Colonel Bland, 30 August, 1777," *The Papers of George Washington*, Vol. 11, 91

[27] Chase and Lengel, eds., "General Washington to John Hancock, 27 August, 1777," *The Papers of George Washington*, Vol. 11, Note 1, 78

Captain Lee gained acclaim when he and his troop returned from a patrol with 24 prisoners.[28] General Washington noted Lee's accomplishment in a letter to John Hancock, the president of Congress. Within a week, Lee's total of captured enemy prisoners surpassed forty.[29]

Lee's success in the field was probably a factor behind General Washington's intervention in an apparent dispute between Lee and his commander, Colonel Bland. General Washington wrote to Bland in late August and instructed him to send all of Captain Lee's dragoons, who Bland apparently had scattered into several detachments, back to Lee. To the chagrin of Bland, General Washington included strong praise for Captain Lee.

> *You will be pleased to send all Capt. Lee's troop that remain behind with the regiment to join the detachment now with him and you will also return immediately to him, whatever men he may send with prisoners, or on any other errand. He is so enterprising and useful an officer that I should wish him not to be straitened for the want of men.*[30]

Although General Washington clearly supported Captain Lee in this instance, he had little time to become embroiled in the endless squabbles of his officers. Reports from Captain Lee and other patrols indicated that General Howe's army was finally about to march.[31] In anticipation, Washington organized a special light infantry corps of 700 men to serve as

---

[28] Chase and Lengel, eds., "General Washington to John Hancock, 30 August, 1777," *The Papers of George Washington*, Vol. 11, 93

[29] Showman, ed., "General Greene to Jacob Greene, 31 August, 1777," *The Papers of General Nathanael Greene*, Vol. 2, 149

[30] Chase and Lengel, eds., "General Washington to Colonel Bland, August 1777," *The Papers of George Washington*, Vol. 11, 104

[31] Showman, "Levi Hollingsworth to General Greene, 2 September, 1777," *The Papers of General Nathanael Greene*, Vol. 2, 152

an advanced guard of the army.[32] This infantry corps, made up of detachments from the army at large, was commanded by General William Maxwell of New Jersey and was supported by hundreds of local militia and the bulk of the Virginia light horse.[33] General Maxwell received orders that were similar to Colonel Bland's, namely to keep a close eye on the enemy.[34] Maxwell's corps was also expected to resist the advance of the enemy. This is exactly what they did when General Howe's army approached on the morning of September 3rd.

Cooches Bridge

Aware that his light corps was significantly outnumbered, General Maxwell determined to harass and delay Howe's advance but give ground when pressed. Maxwell instructed his men to use the natural cover of the wooded terrain to ambush the enemy as they advanced along the road to Iron Hill and Cooches Bridge. When pressed, the Americans would retreat and reform further down the road to strike the enemy again. Every tree, thicket, and rock along the enemy march route was a possible firing position and at least one British officer, Captain John Montresor, commented on the ambush prone terrain.

> *The Country is close – the woods within shot of the road, frequently in front and flank and in projecting points towards the Road.*[35]

---

[32] Chase and Lengel, eds., "General Orders, 28 August, 1777," *The Papers of George Washington*, Vol. 11, 82

[33] Henry Steele Commager and Richard B. Morris, ed. "Captain Walter Steward to General Gates, 2 September, 1777," *The Spirit of Seventy-Six*, (New York: Castle Books, 2002), 610
Originally published in 1958

[34] Chase and Lengel, eds., "General Washington to General Maxwell, 3 September, 1777," *The Papers of George Washington*, Vol. 11, 140

[35] "Journal of Captain John Montresor, 3 September, 1777," *The Pennsylvania Magazine of History and Biography*, Vol. 5, (Philadelphia: The Historical Society of Pennsylvania, 1881), 412

Captain Johann Ewald of the German jagers (riflemen) led the British advance with six mounted dragoons. As they slowly rode ahead of the column, shots erupted from a nearby thicket. Ewald recalled,

> *I...had not gone a hundred paces from the advance guard when I received fire from a hedge, through which these six men* [the dragoons] *were all either killed or wounded. My horse, which normally was well used to fire, reared so high several times that I expected it would throw me. I cried out, "Foot jagers forward!" and advanced with them to the area from which the fire was coming... At this moment I ran into another enemy party with which I became heavily engaged.*[36]

The Americans gradually fell back, pursued by the jagers. Captain Montresor noted that, "*a Continued Smart irregular fire* [ensued] *for near two miles.*"[37] The engagement extended into the afternoon and culminated on heavily wooded Iron Hill and at Cooches Bridge. General Howe ordered his advance guard to drive the Americans from the hill and once again Captain Ewald boldly led the advance. He proudly recalled that,

> *The charge was sounded, and the enemy was attacked so severely and with such spirit by the jagers that we became masters of the mountain after a seven hour engagement.... The majority of the jagers came to close quarters with the enemy, and the hunting sword was used as much as the rifle....*[38]

By mid-afternoon General Howe's troops had pushed the Americans from the hill and bridge. Maxwell's light corps

---

[36] Ewald, 77
[37] Montresor Journal, 412
[38] Ewald, 77

withdrew towards General Washington and the main army south of Wilmington, Delaware. Despite the heavy fighting, casualties were light, estimated at 25 to 50 men for each side.[39]

It does not appear that Captain Lee or many of his fellow cavalrymen participated in the fighting at Cooches Bridge and Iron Hill. The terrain was not conducive for cavalry and there are few references to American horsemen in any part of the battle. It appears instead, that most of the American cavalry spent the day shadowing a large enemy detachment under General Knyphausen in the vicinity of Cecil Court House, across the Elk River.[40]

General Washington expected Howe to resume his march towards Philadelphia the following day and he prepared his army for battle, but General Howe remained stationary for nearly a week, baffling the American commander once again. When Howe finally did advance, he moved around the Americans, not at them. Early in the morning of September 8th, the British broke camp and marched northward, effectively turning Washington's right flank and forcing the Americans to shift northward themselves. Washington had already posted the bulk of his cavalry on his right flank, now he shifted the army in that direction.[41] His destination was Chads Ford, Pennsylvania, a key crossing point over Brandywine Creek. Seven miles to the west, at Kennett Square, General Howe and his army prepared to confront the Americans and settle the conflict once and for all.

---

[39] Michael Cecere, *They Are Indeed a Very Useful Corps: American Riflemen in the Revolutionary War*, (Westminster, MD: Heritage Books, 2006), Note 12, 135

[40] Chase and Lengel, eds., "General Washington to John Hancock, 3 September, 1777," *The Papers of George Washington*, Vol. 11, 136

[41] Chase and Lengel, eds., "General Orders, 7 September, 1777," *The Papers of George Washington*, Vol. 11, 168

Brandywine

Brandywine Creek did not present a significant obstacle to General Howe and his army. Running north to south, the average width of the creek was about 100 yards and although the water level was chest high or higher in many places, there were numerous shallow spots that one could ford. Steep hills bordered both sides of the creek. On the west bank the hills extended up to the creek but on the east bank a flat flood plain a few hundred yards wide separated the creek from the hills.

General Washington posted his army in the hills overlooking the flood plain and creek. He centered his line at Chads Ford, and placed troops north and south of this position. Pennsylvania militia under General John Armstrong guarded the rugged terrain south of Chads Ford while continental troops under General Nathanael Greene and General Anthony Wayne guarded Chads Ford itself. The ground north of Chads Ford was guarded by continental troops under General John Sullivan. A division of continental troops under General William Alexander (Lord Stirling) and another division under General Adam Stephen was posted in reserve near Chads Ford and General Maxwell's advance corps was posted across the creek to reconnoiter and give adequate warning of the enemy's approach.

Colonel Bland's Virginia light cavalry also conducted reconnaissance of the enemy through frequent patrols across the creek. Although General Washington was assured by locals that there were no suitable fords within twelve miles of General Sullivan's right flank, Colonel Bland was ordered to patrol north of Sullivan's right flank.[42]

General Howe commenced his advance towards Chads Ford around dawn on September 11th. He split his 18,000 man

---

[42] Chase and Lengel, eds., "Editorial Note," *The Papers of George Washington*, Vol. 11, 189

army in two.[43] General Wilhelm Knyphausen, an experienced Prussian commander, led approximately one third of Howe's army eastward, directly towards Chads Ford.[44] They soon ran into General Maxwell's advance corps who conducted a fighting withdrawal all the way back to Chads Ford.

General Howe's plan called for Knyphausen to keep Washington's army occupied at Chads Ford while he led the bulk of his army on a long march around the right flank of Washington's line. This move, which Howe had successfully employed at Long Island a year earlier, threatened to crush the Americans in a pincer movement. It also involved some risk for Howe, for by dividing his army, General Howe invited an attack by General Washington upon Knyphausen's smaller force.

In the opening hours of the British operation it appeared to the Americans that the British were mounting a simple frontal assault. By mid-morning, General Maxwell's light corps was pushed back across Brandywine Creek and the main American line braced for action. Knyphausen's troops, however, did not advance across the creek and a tense lull settled along Brandywine Creek.

General Washington began to suspect that Knyphausen's advance to Chads Ford was only a feint. This conclusion was supported in mid-morning by a report from General Sullivan of enemy troop movement northwest of Sullivan's position (across Brandywine Creek to Sullivan's front and right flank). General Washington saw an opportunity to turn the table on Howe and strike his divided force and was in the process of launching an attack on Knyphausen when a new report reached headquarters that generated doubt that Howe had in fact divided his army.

---

[43] Thomas J. McGuire, *The Philadelphia Campaign*, Vol. 1 (Stackpole Books, 2006), 172
[44] Ibid., 173

Brandywine

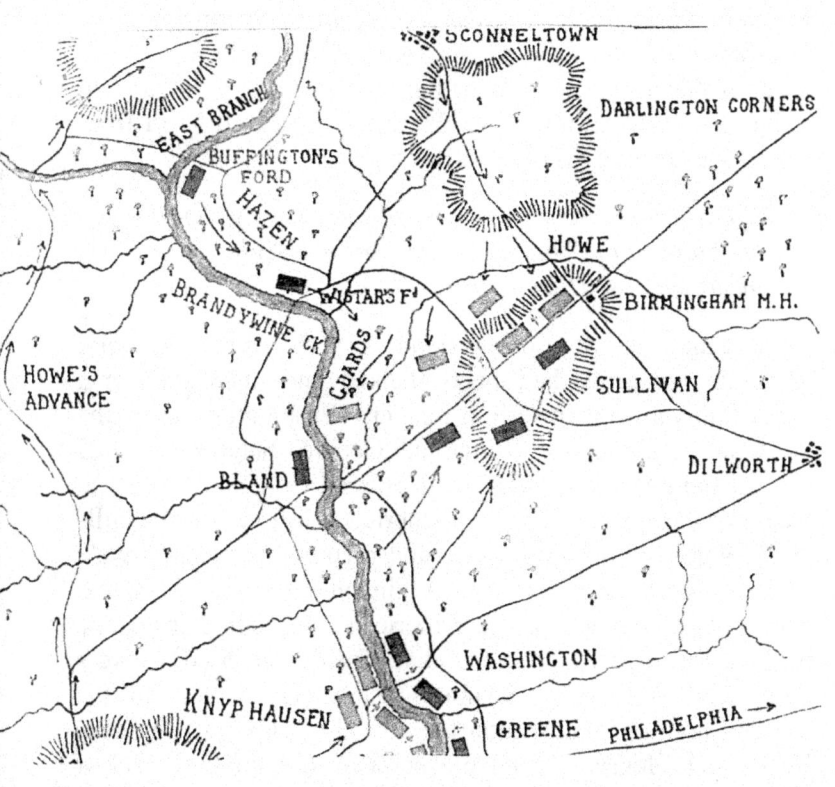

Courtesy of BritishBattles.com

Frustrated by these conflicting reports, General Washington, *"bitterly lament[ed] that Coll. Bland had not sent him any information at all, & that the accounts he had received from others were of a contradictory nature."*[45] As a result of this uncertainty, Washington called off his attack and waited for more intelligence. He finally received word from Bland, via General Sullivan, around 2 p.m.

> *I have discovered a party of the Enemy on the Heights...about half a mile to the Right of the meeting house (Birmingham).*[46]

Once again, as at Long Island a year earlier, General Washington's army had been flanked and his right wing, under Sullivan, was in serious danger. Washington rushed the divisions of General Stirling and General Stephen towards Birmingham Meeting House and ordered General Sullivan to redeploy there as well. The Americans formed on the hills overlooking the meeting house and prepared to meet Howe's attack. One lone Virginia regiment, (the 3rd Virginia) deployed in advance of the American line behind the stone wall of the meeting house and alongside the road Howe's army advanced on. An intense skirmish erupted on the meeting house grounds upon the arrival of Howe's vanguard. Years later, Henry Lee made a point to describe the stand of the 3rd Virginians in his memoirs.

> *Cut off from cooperation...*[the 3rd Virginia] *bravely sustained itself against superior numbers, never yielding one inch of ground, and expending thirty rounds a man, in forty-five minutes.*[47]

---

[45] Chase and Lengel, eds., "Editorial Note, quote from Charles Coteworth Pinckney," *The Papers of George Washington*, Vol. 11, 190
[46] Chase and Lengel, eds., "Colonel Bland to General Washington, 11 September, 1777," *The Papers of George Washington*, Vol. 11, 198
[47] Lee, 89

American artillery fire supported the besieged Virginians and prompted one British officer to observe that,

> *The trees [were] cracking over ones head. The branches riven by the artillery, the leaves falling as in autumn by the grapeshot.*[48]

The Virginians were eventually forced from the meeting house and rejoined their brigade in the rear. The 3rd Virginia paid a heavy price for their brave stand at Birmingham Meeting House. Nearly half of their officers, thirteen non-commissioned officers, and sixty privates fell in the battle.[49] General Howe's troops swept on to engage the entire American right wing. The sound of the battle carried over to Chads Ford, a few miles away. This was the signal that General Knyphausen was waiting for to resume his assault. He commenced a heavy bombardment of the Americans at Chads Ford and followed it with a frontal assault.

The American army was caught in a pincer movement and Washington's troops gave ground on both fronts. The best they could hope for now was to escape, and Washington struggled to do so as orderly as possible. Reinforcements under General Nathanael Greene reached the American right wing just in time to cover their retreat and prevent a complete rout. Lieutenant James McMichael served in Greene's brigade and described their role in the engagement.

> *We took the front and attacked the enemy at 5:30 p.m., and being engaged with their grand army, we at first were obliged to retreat a few yards and formed in an open field when we fought without giving way on either side until dark. Our ammunition almost expended, firing ceased on both sides, when we*

---

[48] Samuel Smith, *The Battle of Brandywine*, (Monmouth Beach, NJ: Philip Freneau Press, 1976), 17
[49] Lee, 89-90

> *received orders to proceed to Chester...This day for a severe and successive engagement exceeded all I ever saw.*[50]

The situation for the American troops at Chads Ford was equally dire. Weakened by the departure of Greene's brigade and overwhelmed by Knyphausen's troops, the American force at Chads Ford gave ground and withdrew eastward. Nightfall spared them, along with their comrades on the right wing, further loss. Exact casualty numbers are impossible to tabulate, but it appears that the Americans lost more than double the nearly 600 British casualties (in killed, wounded, and captured).[51]

Like the engagement at Cooches Bridge a week earlier, Captain Lee and his fellow cavalrymen apparently played a limited role in the battle of Brandywine. Detachments of dragoons were distributed throughout the American army on the day of the battle and some undoubtedly engaged the enemy, but there is no evidence that Captain Lee commanded or was part of any of these detachments.

Years later, critics of Henry Lee III would accuse him of failing to warn General Washington of Howe's flank march, but if such criticism is deserved by anyone, it is best directed at Colonel Bland. It was Bland after all, who was ordered by Washington to reconnoiter the right flank and therefore the responsibility of detailing patrols to the area fell to him. Why General Howe's troops went undiscovered for as long as they did remains unclear, but little evidence exists to implicate Captain Lee of any wrongdoing or incompetence.

---

[50] James McMichael, "The Diary of Lieutenant James McMichael of the Pennsylvania Line, 1776-1778," *The Pennsylvania Magazine of History and Biography*, Vol. 16, no. 2, 1892, 150

[51] McGuire, 269

A week after the battle, Captain Lee had a near fatal brush with the enemy when he and Lieutenant Colonel Alexander Hamilton, an aide to General Washington, were surprised by a party of British cavalry. Lee and Hamilton, escorted by a handful of Lee's dragoons, had been sent to the Schuylkill River to destroy a large supply of flour and some boats before they fell into enemy hands. Lee recounted what happened in his memoirs.

> *The mill, or mills, stood on the bank of the Schuylkill. Approaching, you descend a long hill leading to a bridge over the mill-race. On the summit of this hill two vedettes were posted; and soon after the party reached the mills, Lieutenant-Colonel Hamilton took possession of a flat-bottomed boat....*[52]

Suddenly, the sound of gunfire from Lee's vedettes announced the approach of the enemy. Lee continued his account.

> *The dragoons were ordered instantly to embark. Of the small party, four with the lieutenant-colonel jumped into the boat, the van of the enemy's horse in full view, pressing down the hill in pursuit of the two vedettes. Captain Lee, with the remaining two, took the decision to regain the bridge, rather than detain the boat. Hamilton was committed to the flood, struggling against a violent current, increased by the recent rains; while Lee put his safety on the speed and soundness of his horse.*[53]

The enemy focused their attention on Lee and the two vedettes, all of who raced towards the bridge. Lee recalled,

---

[52] Lee, 91
[53] Ibid.

The two vedettes preceded Lee as he reached the bridge; and himself with the two dragoons safely passed it, although the enemy's front section emptied their carbines and pistols at the distance of ten or twelve paces.[54]

Lee and his men had narrowly escaped, but the sound of gunfire downriver suggested that Hamilton had not been so fortunate.

Lieutenant Colonel Hamilton had reached the same conclusion about Lee and when Hamilton arrived at headquarters he expressed his fears about Lee's welfare to General Washington. The commander-in-chief and his aide were undoubtedly relieved and amused when a letter from Captain Lee arrived describing his escape and expressing Lee's concern for Hamilton's safety. Lee concluded his recollection of this incident in his memoirs with the comment, *"Thus did fortune smile upon those two young soldiers, already united in friendship, which ceased only with life."*[55]

Fight for Philadelphia

Although Captain Lee and Lieutenant Colonel Hamilton had successfully eluded capture, the same was not true for Philadelphia. The de facto American capital fell to General Howe without a fight on September 26th. General Washington struck back at Howe a week later with a surprise attack against Howe's troops at Germantown, a small village just north of Philadelphia. Washington's attack was initially successful, forcing the British light infantry to retreat, but determined resistance by about 100 British troops barricaded in the Chew House (a stone mansion) coupled with mounting confusion among the American troops caused the attack to break down.

---

[54] Ibid.
[55] Ibid., 92

General Washington tried to account for the cause of the failure in a letter to Congress the day after the battle.

> *The Morning was extremely foggy, which prevented our improving the advantages we gained so well.... This circumstance, by concealing from us the true situation of the Enemy, obliged us to act with more caution and less expedition than we could have wished, and gave the Enemy time to recover from the effects of our first impression; and what was still more unfortunate, it served to keep our different parties in ignorance of each Others movements, and hindered their acting in concert. It also occasioned them to mistake One another for the Enemy, which, I believe, more than anything else contributed to the misfortune which ensued. In the midst of the most promising appearances – when everything gave the most flattering hopes of victory, the Troops began suddenly to retreat; and intirely left the Field in spite of every effort that could be made to rally them.*[56]

Once again, Captain Lee and his troop of dragoons, like most of Washington's cavalry at Germantown, saw little combat. Lee and his dragoons spent the day attached to General Washington.[57] Reflecting on the battle years later in his memoirs, Lee attributed four factors for the American failure at Germantown, poor discipline among Washington's troops, their weak and fatigued condition, a complicated plan of attack, and inexperienced commanders.[58] This blunt, yet fair assessment, hit the mark.

---

[56] Chase and Lengel, ed. "General Washington to John Hancock, 7 October, 1777," *The Papers of George Washington*, Vol. 11, 393-94
[57] Lee, 96
[58] Ibid.

Delaware River Defenses

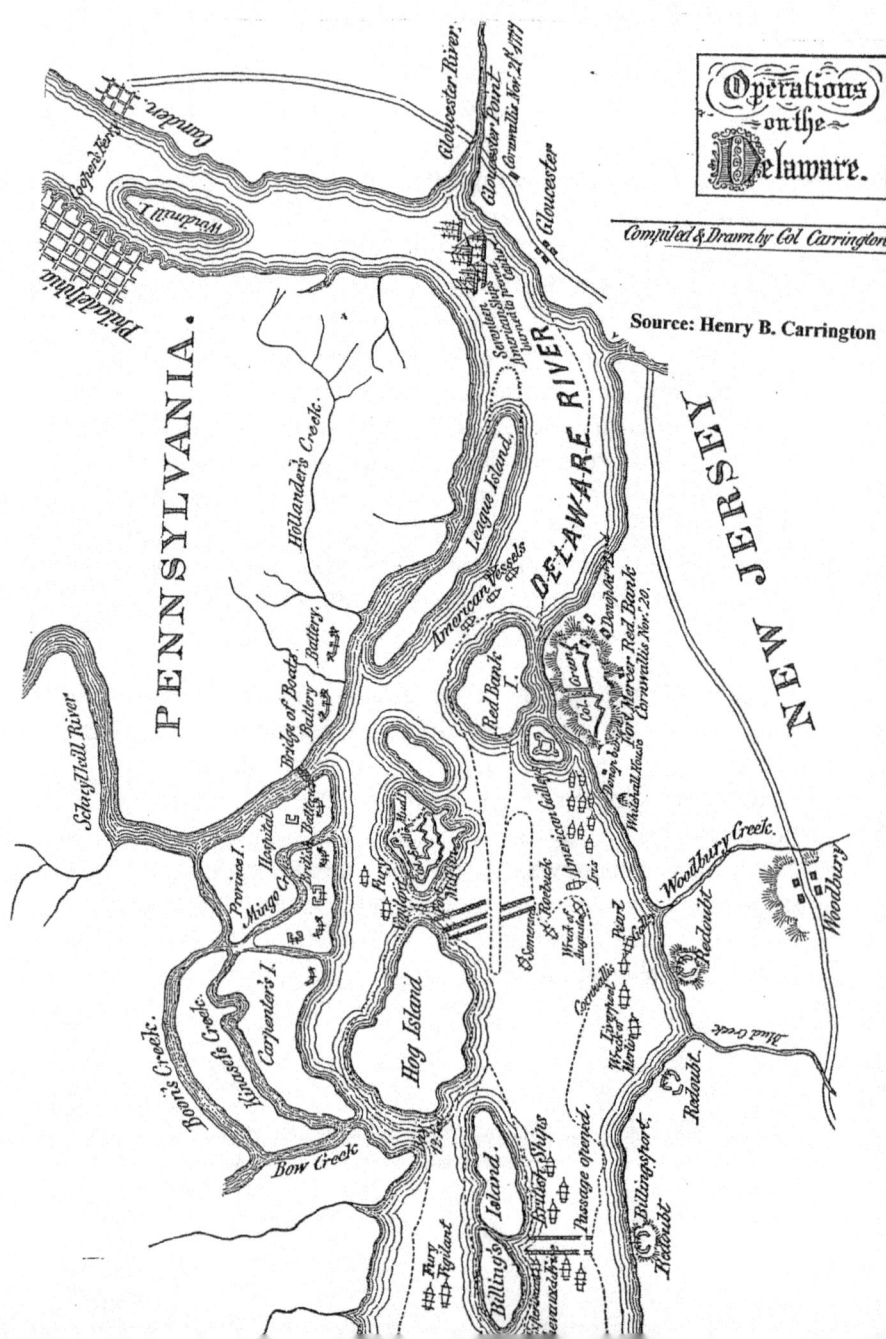

Immediately after the battle, General Washington led his army westward to rest and recover and await reinforcements from New York and Virginia.[59] Determined to closely monitor General Howe and his army in Philadelphia, Washington ordered Captain Lee to patrol up to the outskirts of the city, gather as much intelligence on the enemy as possible, and report directly to Washington on a regular basis.[60]

This assignment marked a shift in Captain Lee's military service. Up to this point, Captain Lee and his troop had served in Colonel Bland's 1st Continental Dragoon Regiment and although they had often been detached from the regiment, their orders and activities generally went through Colonel Bland. From October 1777 onward, however, Captain Lee and his troop were viewed and treated by General Washington as an independent command, receiving their orders directly from the commander-in-chief. Captain Lee undoubtedly welcomed the new arrangement and strove to please General Washington. Lee's reports to Washington included intelligence gathered from numerous patrols as well as interrogations and interviews from deserters, prisoners, and civilians. Lee's first report to Washington was written on October 9th and included positive news. It assured Washington that enemy losses at Germantown were much higher than originally thought.

> Their wounded were brought into the city & laid on straw, strewed in the different streets for that purpose. The most sanguine friends to government gave over all for lost. Consternation & confusion was universal throughout the army & city. The soldiers threw away their guns knap-sacks etc. in the

---

[59] Chase and Lengel, eds., "General Washington to John Hancock, 7 October, 1777," *The Papers of George Washington*, Vol. 11, 417

[60] Chase and Lengel, eds., "Captain Henry Lee Jr. to General Washington, 9 October, 1777," *The Papers of George Washington*, Vol. 11, 463-64

> *flight, their artillery ran in on each other. Ruin appeared to be their inevitable lot.... The enemys pride is lowered, their own fears alarm their camp every night; they certainly must fall an easier sacrifice on another trial.*[61]

A week later, Lee asserted to Washington that there was no need to fear an enemy attack. "*On the contrary, they are in constant apprehension of an attack from your Excelly.*"[62] Captain Lee also warned Washington about the presence of a possible double agent in camp.[63] Near the end of October, General Washington sent Lee and his dragoons across the Schuylkill River to help sever the flow of supplies reaching Howe's army in Philadelphia by land.

Although General Howe's capture of Philadelphia was a serious blow to the American cause, his grip on the city was threatened by the inability of the British navy to overcome the river defenses and obstructions erected by the Americans downriver. These impediments to navigation on the Delaware River were created a year earlier and included underwater obstacles called chevaux-de-frize (large iron tipped poles submerged just below the waterline to rip holes in the bottom of any vessel that passed over them). Sunken ships in the channel and two strong forts, Fort Mifflin on the Pennsylvania shore and Fort Mercer on the New Jersey shore, also blocked passage of the river. The forts were the key to American success and if they could be held and maintained long enough, General Howe might be compelled to abandon Philadelphia due to a lack of provisions and supplies.

---

[61] Ibid.

[62] Chase and Lengel, eds., "Captain Henry Lee Jr. to General Washington, 19 October, 1777," *The Papers of George Washington*, Vol. 11, 545

[63] Ibid. Note: George Spangler, the person identified by Lee as a possible spy, was executed in August 1778 as a spy.

While the British focused their efforts after Germantown against the two American forts, they also developed a tenuous supply route that bypassed the forts over land. Although this laborious method did provide Howe's army with vital supplies, it was not sufficient to maintain the army in Philadelphia over the winter.

General Washington hoped to starve the British out of the city before winter and that meant holding the river forts and cutting off supplies by land. In late October, Captain Lee and his dragoons were sent south, across the Schuylkill River to help achieve this. After observing the British supply route, Lee informed General Washington that the key to halting it was an island manned by 500 enemy troops.

> *The possession of this post* [Carpenter Island] *secures a constant & ready supply of provision. It is brought up by water, from the fleet off Chester, reposited under cover of the ships...& then conducted thro' Carpenters island to the new lower ferry; so on to Philada. If this communication is not interrupted, supplies of provisions, will be as abundant, as if the fleet lay off the wharfs of the city.*[64]

Although Captain Lee and the rest of Washington's army were powerless to halt British operations on Carpenter's Island, Lee did conduct a series of patrols and raids in Delaware to discourage and disrupt commerce between the local inhabitants and the enemy. On one such patrol, Lee and twelve dragoons routed an enemy foraging party, capturing nine men.[65] Captain Lee's efforts in early November gained him recognition from General Washington.

---

[64] Frank E. Grizzard Jr. and David R. Hoth, eds., "Captain Henry Lee Jr. to General Washington, 31 October, 1777," *The Papers of George Washington*, Vol. 12, 67

[65] Grizzard Jr. and Hoth, eds., "Captain Henry Lee Jr. to General Washington, 3 November, 1777," *The Papers of George Washington*, Vol. 12, 104-105

The General desires Capt. Craig, Capt. Lee, and the other officers who have distinguished themselves, will accept his cordial thanks, for the enterprise, spirit and bravery they have exhibited in harassing, and making captives of the enemy.[66]

Despite the fall of Fort Mifflin on November 15th (after a heroic stand by its garrison) Lee continued to patrol the west bank of the Delaware River. The British turned their attention to Fort Mercer on the New Jersey side of the river. Determined to keep possession of Fort Mercer, General Washington sent a large detachment under General Greene to reinforce the garrison. Greene arrived too late to prevent the evacuation of Fort Mercer, but he remained determined to challenge British operations in New Jersey. To do so, General Greene requested cavalry support and General Washington responded by ordering Captain Lee to join Greene.[67] Lee's men had an impact even before they reported to General Greene by capturing nine prisoners on their way to his headquarters.[68]

On the same day that Captain Lee reported to Greene, General Washington ordered Greene to return with his detachment to the main army at Whitemarsh, Pennsylvania. Washington was worried that General Howe planned to attack his divided army and he wanted to consolidate his force north of Philadelphia.[69] Greene led most of his detachment back to Pennsylvania, but he left Captain Lee's troop of cavalry as

---

[66] Gizzard Jr. and Hoth, eds., "General Orders, 9 October, 1777," *The Papers of George Washington*, Vol. 12, 177

[67] Gizzard Jr. and Hoth, eds., "General Washington to General Greene, 22 November, 1777," *The Papers of George Washington*, Vol. 12, 349-350

[68] Showman, ed., "General Greene to General Washington, 25 November, 1777," *The Papers of General Nathanael Greene*, Vol. 2, 216

[69] Showman, ed., "General Washington to General Greene, 25 November, 1777," *The Papers of General Nathanael Greene*, Vol. 2, 217

well as the remnants of Colonel Daniel Morgan's rifle corps in New Jersey to bolster the spirits of the New Jersey militia.

> *I have left the rifle Corps at Haddenfield and Capt Lees troops of light Horse to encourage the Militia and awe the enemy, to prevent their coming out in small parties.*[70]

Colonel Morgan and his rifle corps had recently arrived from New York where they played a decisive role in defeating British General John Burgoyne's 7,000 man army at Saratoga. Morgan's rifle corps was created earlier in the summer with 500 select riflemen. Months of hard service in New Jersey and New York had reduced the rifle corps to less than 200 men, but they were still viewed with awe by many people.[71] General Lafayette was one such person. He commanded a detachment that included some of the riflemen in a large skirmish in late November and was reportedly charmed by their spirited behavior.[72] LaFayette observed that,

> *I never saw men so merry, so spirited, so desirous to go on to the enemy what ever forces they could have as the little party was in this little fight. I found the riflemen above even their reputation.*[73]

Captain Lee remained in New Jersey with his troop of dragoons into early December, reporting to General Washington on December 3rd that, *"We have been active in executing* [General Greene's] *directions,"* namely to, *"assist in*

---

[70] Showman, ed., "General Greene to General Washington, 27 November, 1777," *The Papers of General Nathanael Greene*, Vol. 2, 222

[71] Gizzard Jr. and Hoth, eds., "General Washington to General Greene, 22 November, 1777," *The Papers of George Washington*, Vol. 12, 349-350

[72] Showman, ed., "General Greene to General Washington, 26 November, 1777," *The Papers of General Nathanael Greene*, Vol. 2, 218

[73] Gizzard Jr. and Hoth, eds., "General LaFayette to General Washington, 26 November, 1777," *The Papers of George Washington*, Vol. 12, 418-419

protecting the inhabitants from the depredations of the enemy."[74]

Within days of this letter, Lee and his men, along with Colonel Morgan's rifle corps, were back in Pennsylvania. The riflemen joined Washington's main army at Whitemarsh where they skirmished with the advance guard of General Howe's army in what became an abortive attack by Howe. Lee and his troop of dragoons were ordered to patrol in the vicinity of Darby, a small town five miles southwest of Philadelphia.[75] They spent nearly two weeks in the area before they made their way north to join the army at Valley Forge.

---

[74] Gizzard Jr. and Hoth, eds., "Captain Henry Lee Jr. to General Washington, 3 December, 1777," *The Papers of George Washington*, Vol. 12, 528

[75] John W. Hartmann, *The American Partisan: Henry Lee and the Struggle for Independence, 1776-1780*, (Burd Street Press, 2000), 52

Philadelphia Area

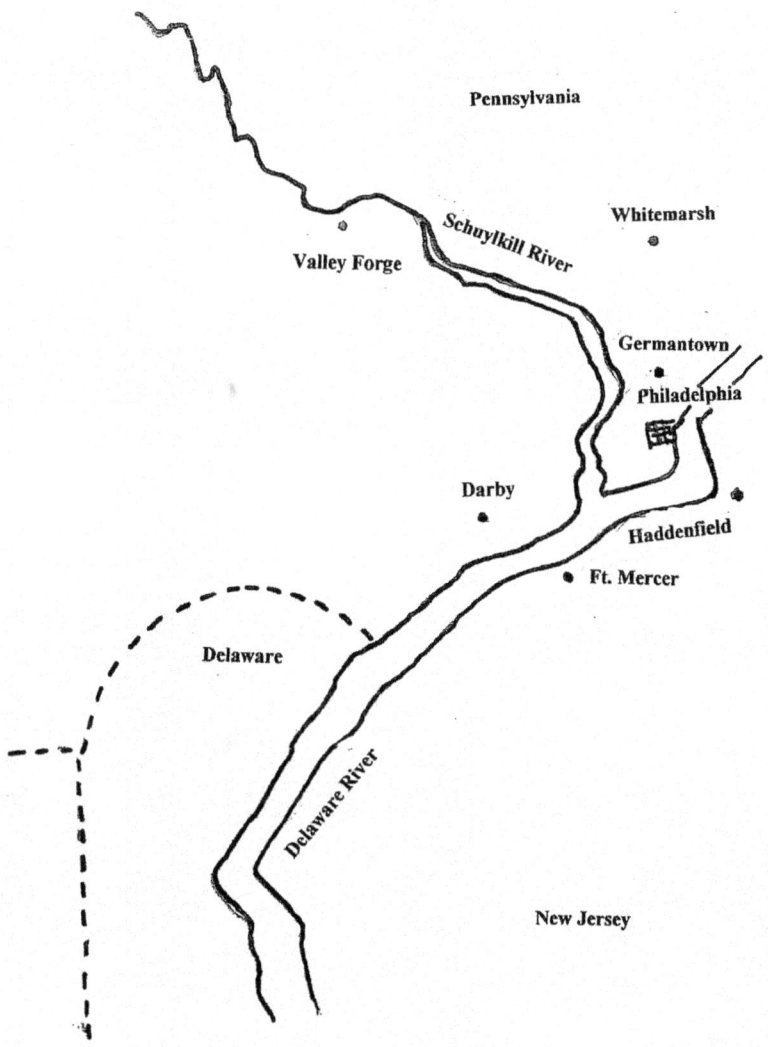

Chapter Three

"Captn Lee's Vigilance baffled the Enemy's designs..."

The American army had marched into Valley Forge on December 19th to establish their winter encampment. General Washington chose to spend the winter there because of its proximity to Philadelphia (approximately 30 miles), its natural defensive features, and his expectation that adequate provision could be obtained for his troops. An encampment only one day's march from Philadelphia also sent a defiant message that the American army, despite the disappointments of the fall, still possessed some fight.

To emphasize this last point and screen the main army while they constructed winter huts and earthworks, General Washington ordered light parties to patrol south of Valley Forge and engage enemy patrols and foraging parties whenever possible. Captain Lee and his troop of dragoons were one of the light parties assigned to this duty.

Working in the vicinity of Radnor Meeting House, roughly seven miles southeast of Valley Forge, Lee's patrols extended south and frequently engaged enemy troops within a few miles of Philadelphia. In one incident, one of Captain Lee's dragoons was captured, but was then rescued by a party of Colonel Morgan's riflemen. Captain Lee described the incident to General Washington.

> *On hearing of the enemys excursion I immediately left camp, & moved down towards Darby. Early this morning we sat out on the partisan business; having fully reconnoitered the enemys disposition, whom we found posted in force...I divided my Troop. Lt. Lindsay with Major Clark whom we accidentally met with took the route towards Chester, while myself with the other party visited the right of their encampment.*[1]

Lee reported that, "*nothing of consequence*," happened to his party, but one of the dragoons with Lieutenant Lindsay was captured by a party of enemy horsemen.[2] Lindsay sought the assistance of Lieutenant Colonel Richard Butler, who was nearby with a large detachment of Morgan's riflemen, and they struck back, capturing ten of the enemy horsemen and freeing Lee's captive dragoon.[3] Lee's troop further benefitted from the affair when they were allowed to keep some of the gear of the captured enemy dragoons.[4]

Captain Lee and his troop continued their patrols outside of Valley Forge into the new year and unlike the bulk of the American cavalry, which spent the winter safely in Trenton, Lee's troop spent most of January posted at a stone farmhouse (Scott's Farm) near Radnor Meeting House.[5] On January 4th, Captain Lee shared his thoughts with General Washington on what was needed to properly reconnoiter and secure the area around Radnor.

---

[1] Gizzard Jr. and Hoth, eds., "Captain Henry Lee to General Washington, 23 December, 1777," *The Papers of George Washington*, Vol. 12, 689
[2] Ibid.
[3] Ibid.
[4] Grizzard Jr. and Hoth, eds., "General Stirling to General Washington, 24 December, 1777," *The Papers of George Washington*, Vol. 12, 696-697
[5] Edward G. Lengel, ed., "General Washington to General Casimir Pulaski, 31 December, 1777," *The Papers of George Washington*, Vol. 13, 89

> *Agreeable to your Excellency's direction I have informed myself minutely with the country in the vicinity of Radnor meeting-house. To effect the object of your Excellency's wishes, vizt. security to the camp: I conceive it absolutely necessary to establish two posts of horse. The one to...patrole one mile, more or less, in advance of the advanced centinal. The other to be fixed near Newtown-square...to patrole the square-tavern. This last place, is without any guard.... It appears to me, that the post established, or to be established at the meeting-house is very far from being secure, unless great attention is paid to the square, at which place the three roads Hartford, Darby & Chester all unite.*[6]

Captain Lee explained that two troops of cavalry were needed to secure the area at night, but in the daytime, one troop of horse could handle the job. Lee went on to state the necessity of impressing provisions from local farmers and he offered to engage in the business of intelligence gathering by serving as a contact for American spies in and around Philadelphia.[7]

With nearly all of the army's cavalry posted in Trenton, there were few dragoons left at Valley Forge to send to Lee so his request for additional dragoons was ignored. Lee had to rely on Colonel Morgan's riflemen at Radnor Meeting House for support and Morgan relied on Lee's dragoons to serve as videttes on Morgan's thin piquet line.

Undoubtedly frustrated by the difficult situation he found himself in, Captain Lee was pleasantly surprised to learn in mid-January that he and his officers had gained some notoriety in the *New Jersey Gazette*. The newspaper credited Captain Lee and his troop of dragoons with capturing over 125 enemy

---

[6] Lengel, ed., "Captain Henry Lee to General Washington, 4 January, 1777," *The Papers of George Washington*, Vol. 13, 141-142
[7] Ibid.

prisoners since the summer and called for the service of Lee and his officers to be duly rewarded.

> *A troop of dragoons in Bland's regiment, seldom having more than 25 men and horses fit for duty, has since the first of August last, taken 125 British and Hessian privates, besides four commissioned officers, with the loss of only one horse. This Gallant Corps is under the command of Captain Lee, Lt. Lindsay and Cornet Peyton whose merit and services it is hoped will not be passed unnoticed or unrewarded.*[8]

Raid on Scott's Farm

Lee's notoriety extended to the enemy, who sent a large cavalry detachment to Scott's Farm to seize Lee late in the evening of January 18th. An aide to British General William Howe noted the effort in his diary.

> *At 11 o'clock at night, 40 dragoons were detached by a long roundabout way to seize a rebel dragoon captain by the name of Lee, who has alarmed us quite often by his boldness....*[9]

Captain Johann Ewald recorded a similar observation in his journal but doubled the size of the British force.

> *Today the English Major Crewe was sent out with eighty horsemen to surprise the partisan Captain Lee, who stood with forty horse on this side of Valley Forge and constantly alarmed our outposts.*[10]

---

[8] *New Jersey Gazette*, 14 January, 1778
[9] Lengel, ed., "Captain Muenchhausen's Diary," *The Papers of George Washington*, Vol. 13, 292-293
[10] Ewald, 121

The British rationale for the attack, to seize a bold American officer who had become a nuisance, was also mentioned in at least one American account of the affair. The *New Jersey Gazette* attributed the attack to General Howe's,

> *Longing to rob the Americans of this gallant young officer, whose attention in observing his motions, and address in surprising his parties perplexed him so much the last campaign.*[11]

Captain Lee was posted at Scott's Farm, about 16 miles west of Philadelphia, with Lieutenant Lindsay and a handful of dragoons. Major John Jameson of the 1st Continental Dragoons (Bland's regiment) was also at the farm paying Lee a visit. The bulk of Lee's troop was out on patrol in small parties and the handful of officers and men left at the farm were sheltered in a strong stone farmhouse.

Captain Lee described the opening of the engagement in a letter to General Washington after the affair.

> *About day break* [the enemy] *appeared, we were immediately alarm'd, & manned the doors & windows.*[12]

According to an account of the engagement in the *New Jersey Gazette*, Lee's men,

> *Scarcely had time to bolt the doors before* [the enemy] *began a smart firing into the windows, and demanded the immediate surrender of the house....* [Captain Lee refused to surrender and his men] *returned the fire from the windows with spirit, and,*

---

[11] Moore, "New Jersey Gazette, 28 January, 1778," *Diary of the American Revolution*, Vol. 2, 10

[12] Lengel, ed., "Captain Henry Lee Jr. to General Washington, 20 January, 1778," *The Papers of George Washington*, Vol. 13, 292

> by showing themselves at different places, made as great an appearance of numbers as possible."[13]

The British dragoons repeatedly tried to storm the house and were driven back each time. Frustrated, some of them began plundering the outbuildings which prompted Captain Lee to brazenly taunt the enemy commander.

> *Comrade, shame on you, that you don't have your men under better discipline. Come a little closer, we will soon manage it together!*[14]

After about a half hour the British gave up and withdrew, suffering eight casualties.[15] Although four of Lee's dragoons on patrol were captured by the British on their way back to Philadelphia and another was captured at the farm when he tried to flee, Lee and his men emerged victorious from the engagement and were showered with praise from the public. General Washington publically acknowledged Lee's brave stand, thanking Lee and his men in the general orders.

> *The Commander in Chief returns his warmest thanks to Captn Lee & the Officers & men of his Troop for the Victory which by their superior Bravery and Address they gain'd over a party of the Enemys dragoons, who trusting in their numbers – and concealing their march by a circuitous road attempted to surprise them in their quarters. He has the satisfaction of informing the Army that Captn Lee's Vigilance baffled the Enemy's designs by judiciously posting his men in his quarters, although he had not a sufficient number to allow one for each*

---

[13] Moore, "New Jersey Gazette, 28 January, 1778," *Diary of the American Revolution*, Vol. 2, 10
[14] Ewald, 121
[15] Lengel, ed., "Captain Henry Lee Jr. to General Washington, 20 January, 1778," *The Papers of George Washington*, Vol. 13, 293

> *window, he obliged the [enemy] disgracefully to retire after repeated but fruitless attempts to force their way into the house.*[16]

General Washington followed this glowing public praise for Lee with a private letter to the young captain that hinted at a reward to come.

> *Altho I have given you my thanks in the general Orders of this day for the late instance of your gallant behavior I cannot resist the Inclination I feel to repeat them again in this manner. I needed no fresh proof of your merit, to bear you in remembrance – I waited only for the proper time and season to shew it – these I hope are not far off.... Offer my sincere thanks to the whole of your gallant party and assure them that no one felt pleasure more sensibly, or rejoiced more sincerely for yours & their escape than Yr, Affectionate.*
>
>

G. Washington[17]

---

[16] Lengel, ed., "General Orders, 20 January, 1778," *The Papers of George Washington*, Vol. 13, 286-287

[17] Lengel, ed., "General Washington to Captain Lee, 20 January, 1778," *The Papers of George Washington*, Vol. 13, 294

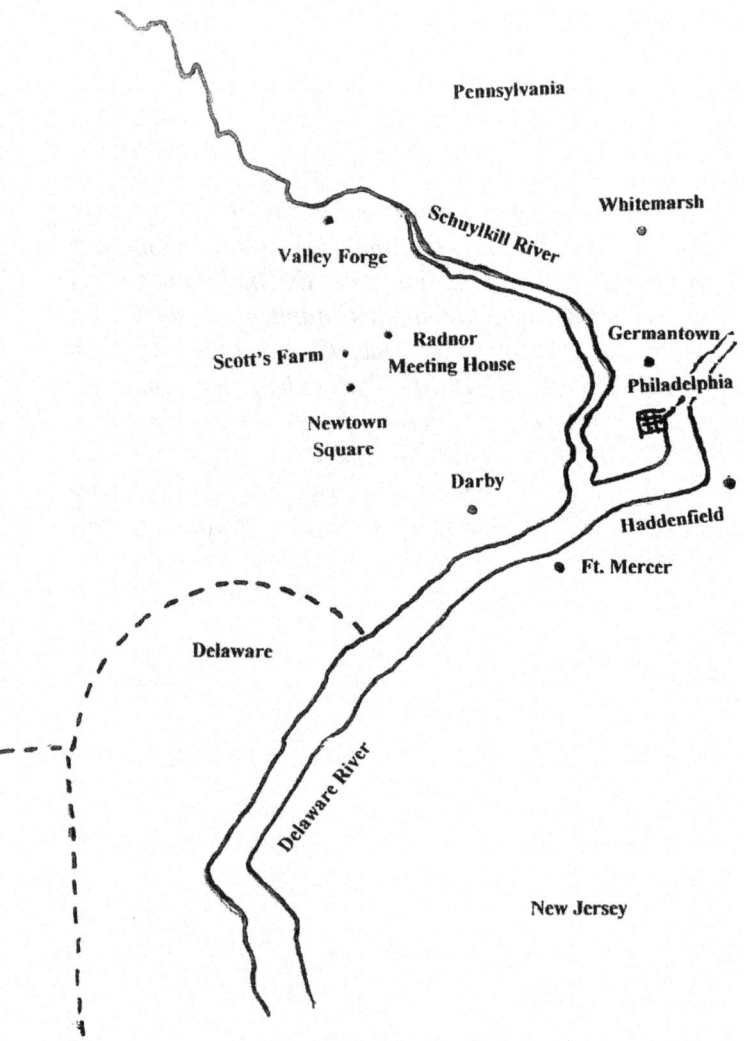

Valley Forge

As evidenced by the American reaction to the skirmish at Scott's Farm, Lee's bold stand was welcome news to an encampment struggling to feed itself at Valley Forge. Although most of the men were in huts by mid-January, adequate provisions and clothing for the troops remained a serious problem for the army. Washington had struggled with a shortage of supplies from the start of the encampment at Valley Forge and he warned Congress then of dire consequences if the shortage was not addressed.

> *I do not know from what cause this alarming deficiency, or rather total failure of Supplies arises: But unless more vigorous exertions and better regulations take place in that line and immediately, This Army must dissolve.*[18]

The lack of supplies contributed to a startling drop in the number of men fit for duty. In many units, two thirds of the men were listed as unfit for duty due to illness or the lack of proper clothing.[19] One way Washington combated this problem, (and encouraged re-enlistments) was to grant winter furloughs to those who agreed to re-enlist. Hundreds returned to their homes on furlough after promising to return in the spring for service. Those that remained suffered through shortages that peaked in February. Major Alexander Scammell of Pennsylvania painted a bleak picture at the height of the crisis in February.

---

[18] Chase, ed., General Washington to Henry Laurens, 22 December, 1777," *The Papers of George Washington*, Vol. 12, 667

[19] Charles H. Lesser, ed., *The Sinews of Independence: Monthly Strength Reports of the Continental Army*, (Chicago: The University of Chicago Press, 1976), 58

> *A moments Opportunity presents of telling you our Distress in Camp has been infinite.... In all the Scenes since I have been in the army, want of provisions these ten Days past, has been the most distressing, [a] great part of our Troops 7 Days with only half a pound of Pork during the whole time – Our poor brave Soldiers living upon bread & water & naked exhibited a Sight exceedingly affecting to the Officers.*[20]

In one of his many appeals for assistance, General Washington shared a similar observation about the troops with George Clinton of New York.

> *For some days past, there has been little less, than a famine in camp. A part of the army has been a week, without any kind of flesh, and the rest three or four days. Naked and starving as they are, we cannot enough admire the incomparable patience and fidelity of the soldiery, that they have not been ere this excited by their sufferings, to a general mutiny and dispersion. Strong symptoms, however, of discontent have appeared in particular instances; and nothing but the most active efforts every where can long avert so shocking a catastrophe.*[21]

Captain Lee and his troop were spared much of the hardship endured in camp (one of the few advantages of being posted in the countryside). They continued to patrol south of Valley Forge but with winter firmly settled over the region, encounters with the enemy dropped significantly. In mid-February, at the height of the supply crisis, Washington turned

---

[20] Joseph Lee Boyle, ed., "Alexander Scammell to Timothy Pickering, 19 February, 1778," *Writings from the Valley Forge Encampment of the Continental Army*, Vol. 2 (Bowie MD: Heritage Books Inc., 2001), 50

[21] Lengel, ed., "General Washington to George Clinton, 16 February, 1778," *The Papers of George Washington*, Vol. 13, 552-553

to Lee and a few other detachments to scour the countryside for provisions. Lee was ordered to lead his dragoons into Delaware and Maryland to, "*exert your utmost endeavours, to hasten to this Army all the flesh provisions, deposited in the Magazines, at those places.*"[22] Lee was empowered to impress (take) whatever wagons he needed and to collect cattle and forage about the countryside after he had forwarded the provisions in the magazines. "*I need say nothing to animate your Zeal on this occasion,*" wrote Washington, "*I am confident you have too just a sense of our necessities to omit any exertion it will possibly be in your Power to make.*"[23]

Captain Lee's time in Delaware and Maryland was productive yet frustrating for the young captain. He and his men managed to collect and forward nearly 100 head of cattle to camp, but had trouble obtaining enough wagons to transport the salted provisions and grain they gathered.[24] Lee was especially frustrated by the degree to which the local inhabitants aided American deserters and informed General Washington that hundreds were in the area.[25] He urged Washington to address the Delaware legislature about halting the practice of aiding deserters and sought authority from the general to, "*apprehend & deliver to the Provost, such persons* [civilians] *as are notoriously guilty of this high crime against the interest of the States.*"[26] This would not be the only time Lee called for firm measures to halt desertions.

The efforts of Captain Lee's troop and the other forage parties helped alleviate the supply crisis at Valley Forge and by March, the army's attention had turned to the new drill being introduced by a volunteer German officer named Baron

---

[22] Lengel, ed., "General Washington to Captain Henry Lee, Jr., 16 February, 1778," *The Papers of George Washington*, Vol. 13, 561-562
[23] Ibid.
[24] Lengel, ed., "Captain Henry Lee Jr. to General Washington, 21 February, 1778," *The Papers of George Washington*, Vol. 13, 632
[25] Ibid.
[26] Ibid.

von Steuben. Lieutenant John Marshall of the 11th Virginia Regiment, (and future chief justice of the Supreme Court) witnessed Steuben's efforts and observed years later that,

> *This gentleman was a real service to the American troops. He established one uniform system of field exercise; and, by his skill and persevering industry, effected important improvements through all ranks of the army during its continuance at Valley Forge.*[27]

While Washington's troops practiced Steuben's new drills, the commander-in-chief moved to add a new officer to his military staff. In late March, Lieutenant Colonel Alexander Hamilton, on behalf of General Washington, asked Captain Lee to join Washington's staff as an aide-de-camp. Lee must have been overwhelmed by the offer; to serve as an aide to General Washington was prestigious and highly coveted. Yet, the young captain shockingly turned Washington down.

Concerned about General Washington's reaction to his refusal, Lee wrote to the commander-in-chief to assure him of,

> *The high sense of gratitude I feel from your Excellency's approbation of my conduct. I assure you, Sir, to deserve a continuance of your Excellency's patronage, will be a stimulus to glory, second to none in power, of the many that operate on my soul....*[28]

Lee also acknowledged the honor and opportunity Washington's offer presented.

---

[27] John Marshall, *The Life of George Washington*, Vol. 2 (Fredericksburg, VA: The Citizens Guild of Washington's Boyhood Home, 1926), 439

[28] David R. Hoth, ed., "Captain Henry Lee Jr. to General Washington, 31 March, 1778," *The Papers of George Washington*, Vol. 14, 368-369

> *Permit me to premise that I am wedded to my sword, and that my secondary object in the present war, is military reputation. To have possessed a post about your Excellency's person is certainly the first recommendation I can bear to posterity, affords a field for military instruction, would lead me into an intimate acquaintance with the politics of the States, and might present more immediate opportunitys of manifesting my high respect and warm attachment for your Excellencys character and person. I know, it would also afford true and unexpected joy to my parents and friends.*[29]

These factors, however, could not overcome the sentiments he held for the cavalry service.

> *I possess a most affectionate friendship for my soldiers, a fraternal love for the two officers who have served with me* [Lieutenant Lindsay and Cornet Peyton] *a zeal for the honor of the Cavalry, and an opinion, that I should render more real service to your Excellency's arms.*[30]

Lee concluded by writing that he was very pleased that his conduct had earned General Washington's approval and that he would cheerfully serve in whatever capacity Washington deemed necessary. It was clear, however, that Captain Lee preferred to remain in the cavalry, and in the cavalry he stayed.

---

[29] Ibid.
[30] Ibid.

Chapter Four

"Capt. Lee's genius particularly adapts him to [this] command"

Captain Lee's decision to forsake General Washington's offer undoubtedly surprised and disappointed the commander-in-chief. Although Washington was not familiar with such rejection, he assured Captain Lee that he understood Lee's sentiment and held no ill will towards him.

> *The undisguised manner in which you express yourself cannot but strengthen my good opinion of you. As the offer on my part was purely the result of a high sense of your merit, and as I would by no means divert you from a Career in which you promised yourself greater happiness, from its affording more frequent opportunities of acquiring Military fame, I entreat you to pursue your own inclinations, as if nothing had passed on this Subject....*[1]

Days after Washington's assurance to Lee, the general wrote to Congress on Captain Lee's behalf.

> *Captain Lee of the light Dragoons and the Officers under his command having uniformly distinguished themselves by a conduct of exemplary zeal, prudence and bravery, I took occasion...to express the high*

---

[1] Hoth, ed., "General Washington to Captain Henry Lee, Jr., 1 April, 1778," *The Papers of George Washington*, Vol. 14, 379

> sense I entertained of their merit, and to assure him, that it should not fail of being properly noticed.... I had it in contemplation at the time...to make him an offer of a place in my family. I have consulted the Committee of Congress upon the Subject, and we were mutually of opinion, that giving Capt. Lee the command of two troops of Horse on the proposed establishment with the Rank of Major, to act as an independent partisan Corps, would be a mode of rewarding him, very advantageous to the Service. Capt. Lee's genius particularly adapts him to a command of this nature, and it will be most agreeable to him, of any station, in which he could be placed.[2]

Congress acted quickly on General Washington's request and passed the following resolution on April 7, 1778.

> Whereas Captain Henry Lee, of the light dragoons, by the whole tenor of his conduct during the last campaign, has proved himself a brave and prudent officer, rendered essential service to his country, and acquired to himself, and the corps he commanded, distinguished honour, and it being the determination of Congress to reward merit,
>
> Resolved, That Captain H. Lee be promoted to the rank of major commandant; that he be empowered to augment his present corps by inlistment to two troops of horse, to act as a separate corps.[3]

---

[2] Hoth, ed., "General Washington to Henry Laurens, 3 April, 1778," *The Papers of George Washington*, Vol. 14, 390-391

[3] Worthington C. Ford, ed., "7 April, 1778," *Journals of the Continental Congress, 1774-1789*, Vol. 10, (Washington, D.C., 1904-37), 315

Congress also promoted William Lindsay to captain of Lee's first troop (his old 5th troop) and Henry Peyton to captain-lieutenant of Lee's second troop.[4]

Major Lee enthusiastically embraced his new command and immediately took measures to build his new partisan corps. Within weeks of his promotion, Congress issued a warrant to Major Lee for $50,000 to purchase horses and accoutrements.[5] Lee also established Charlestown, Maryland as a point of rendezvous and began the process of recruiting new dragoons.[6]

Lee faced a daunting challenge because enthusiasm for the war had ebbed considerably by 1778 so recruits were hard to find. On top of that, Lee was competing with the four other continental cavalry regiments, all of which were in great need for horses, men, and arms. When Congress increased the size of each cavalry troop to 54 troopers on May 27th, the demand for adequate horses and men soared.[7] Lee's challenge increased further the following day when Congress added a third troop of cavalry, under Captain Robert Forsyth of Fredericksburg Virginia to his corps.[8]

Major Lee and his officers scrambled during the summer to raise the necessary men and horses. Lee apparently spent some of this time in Virginia during which he delivered letters from Congress to Governor Thomas Johnson of Maryland and Governor Patrick Henry of Virginia.[9] In July, Lee placed

---

[4] Ibid.
[5] Paul H. Smith, ed. "Henry Laurens to Henry Lee, 28 April, 1778," *Letters of Delegates to Congress*, Vol. 9, 516
[6] *Virginia Gazette*, (Purdie), 10 July, 1778, 3
[7] Ford, ed., "27 May, 1778," *Journals of the Continental Congress*, Vol. 11, 539
[8] Ford, ed., "28 May, 1778," *Journals of the Continental Congress*, Vol. 11, 545 and
John H. Gwathmey, *Historical Register of Virginians in the Revolution*, (Richmond, VA: Dietz Press, 1938), 282
[9] Smith, ed.,"Henry Laurens to Patrick Henry, 27 June, 1778," *Letters of Delegates of Congress*, Vol., 10, 197

announcements in the *Virginia Gazette* instructing his officers and men to, "*repair, without loss of time, to Charlestown, in the State of Maryland, the place of general rendezvous.*"[10] By August, Lee and his corps headed north to New York to rejoin Washington's army. Dressed in new uniforms and mounted on strong, powerful horses, Lee's dragoons projected a martial appearance to everyone they encountered.

> *On Saturday last arrived here, on their way to camp, a large body of cavalry, under the command of the celebrated Major Lee, who has so frequently distinguished himself as a partisan. This corps consists of chosen men, whose courage and activity the Major has tried; and being completely uniformed and extremely well mounted, they made an elegant and martial appearance.*[11]

While Major Lee was undoubtedly pleased by the admiration his new partisan corps received, he was also keenly aware that the military situation had changed considerably during his absence. The biggest development was an alliance between the United States and France. News of this alliance reached America in May and ignited celebrations throughout the states. French military aid, troops, and most importantly, the French navy, would offset Britain's advantage on land and dominance of the sea and perhaps even allow the Americans and their new ally to take the offensive.

The threat of a joint American and French operation prompted General Henry Clinton, the new commander of British forces in America, to abandon Philadelphia and return to New York in June. While thousands of demoralized Tories and a portion of Clinton's army squeezed onto ships and sailed

---

[10] *Virginia Gazette*, (Purdie) 10 July, 1778, 3 and (Dixon & Hunter) 17 July, 1778, 4

[11] *Virginia Gazette*, (Purdie) 21 August, 1778, "Extract of a letter from head quarters, August 4, 1778," 2

down the Delaware River, over 10,000 British troops escorted an enormous baggage train across New Jersey.[12] General Washington cautiously pursued Clinton into New Jersey with approximately the same number of troops.[13] He sent large detachments ahead of his main body to harass Clinton's flanks and ordered the commander of the advance guard, General Charles Lee (who had recently returned to the army after sixteen months of captivity) to strike Clinton's rear guard if possible.

Monmouth

The result was the Battle of Monmouth, an intense day long fight in 100 degree temperatures. General Lee commenced his attack on Clinton's rearguard on the morning of June 28th and initially met with success, but poor communication among his detachments and the arrival of British reinforcements halted Lee's advance and soon his troops were streaming rearward.

General Washington, who rode ahead of the main body of American troops when the battle began, met Lee's fleeing men and took charge of the situation. He rallied a few detachments, which briefly held their ground and slowed the enemy advance. This provided time for Washington's main body to deploy in a strong position along a ridge in the rear. By the afternoon, the battle had developed into an intense artillery duel. Both sides attempted to push the other, but neither gained an advantage, so the battle remained a stalemate into the evening.

Strewn across the fields of Monmouth that night were scores of dead and wounded soldiers. Each side suffered approximately 350 casualties in killed, wounded, and

---

[12] Mark M. Boatner III, *Encyclopedia of the American Revolution*, 3rd ed. Stackpole Books, 1994, 716

[13] Fitzpatrick, ed., "Council of War, 24 June, 1778," *The Writings of George Washington*, Vol. 12, 116

captured, and each side viewed the fight as a victory.[14] General Clinton's baggage train had safely escaped as did his army under cover of darkness late in the evening, but General Washington's troops held the field and had fought the British to a standstill. General Washington made clear who he thought the victor was in the next day's general orders.

> The Commander in Chief congratulates the Army on the Victory obtained over the Arms of his Britanick Majesty yesterday and thanks most sincerely the gallant officers and men who distinguished themselves upon the occasion....[15]

Major Lee deeply regretted that his absence from the army cost him and his corps an opportunity to distinguish themselves at Monmouth. Lee lamented in a letter to his friend, General Anthony Wayne, that, *"I have been almost melancholy by my absence from the army. The name of Monmouth reproaches me to my very soul."*[16] He was eager to win new laurels in battle and upon his return to the army in August was given an opportunity to do so when his corps was attached to General Charles Scott's light infantry detachment.

General Scott was an experienced brigadier general from Virginia who had fought in the battles of Trenton and Princeton, Brandywine and Germantown, and Monmouth. On August 8th, Washington ordered Scott to take command of a new light infantry corps. It was to consist of, *"the best, most hardy and active Marksmen and* [be] *commanded by good Parizan officers."*[17]

---

[14] Boatner III, 725
[15] Fitzpatrick, ed., "General Orders, 29 June 1778," *The Writings of Washington*, Vol. 12, 130
[16] "Major Henry Lee to General Anthony Wayne, 24 August, 1778," Anthony Wayne Papers, (also Hartmann, 76)
[17] David R. Hoth, ed., "General Orders, 8 August, 1778," *The Papers of George Washington*, Vol. 16, 267

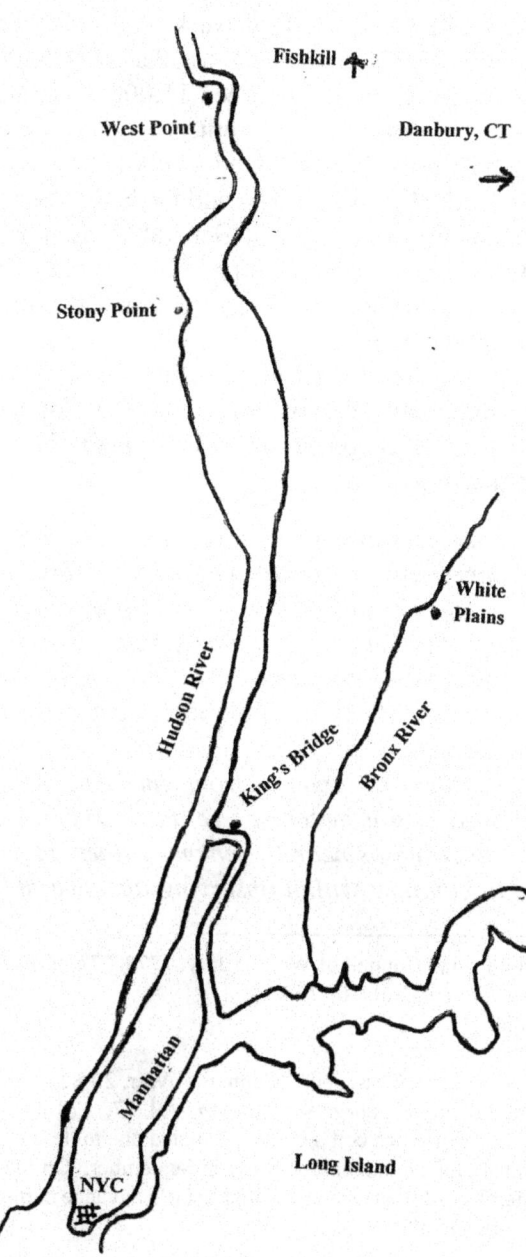

At the time that Washington formed the light corps his army had entered its third week at White Plains, New York, just fifteen miles north of Manhattan Island. Washington estimated that he had between 11,000 – 12,000 men at White Plains. Nearly 5,000 other troops were scattered in detachments in Rhode Island (2,000), New Jersey (1,500) the New York Highlands (900) and the New York frontier (400).[18] Washington estimated that General Clinton had 14,000 troops in the vicinity of New York.[19] As a result, Washington and his officers decided to be cautious and remain at White Plains to await developments.[20]

General Scott's 1,500 man light infantry corps was given the important task of screening the American army and gathering intelligence on enemy activities.[21] Washington instructed Scott to,

> *Take post in front of our camp in such a position as may appear best calculated to preserve the security of your own troops and cover the army from surprise. For the better execution of these purposes you will make yourself master of all the roads leading to the enemies lines. You will keep up a constant succession of scouting parties as large as can possibly be spared from the detachment without harassing it by too severe duty. These parties are to penetrate as near the enemy's lines as possible, and to continue within observing distance at all times....*

---

[18] Hoth, ed., "Council of War, 25 July, 1778," *The Papers of George Washington*, Vol. 16, 163

[19] Ibid.

[20] Ibid.

[21] Philander D. Chase, ed., "Council of War, 29 September, 1778," *The Papers of George Washington*, Vol. 17, 178
Note: The size of Scott's corps fluctuated due to desertions and the arrival of most of the continental cavalry in September. On September 29th, General Washington reported that Scott's advance corps consisted of 1,300 infantry and 400 cavalry.

> *These parties will make you constant reports of their discoveries, and you will give me the earliest and fullest intelligence of all occurrences worthy of notice.*[22]

By the time Lee and his dragoons joined General Scott in early September, Scott's light troops had engaged in a number of skirmishes with the enemy. One engagement in particular decimated a company of 40 Stockbridge Indians serving in Scott's corps. The company was commanded by Captain Abraham Nimham and included Nimham's father, Daniel. Both were respected members of the Stockbridge Indians. Nimham's company and a company of troops under Major John Steward, were ambushed by a large enemy party of horse and infantry. Captain Johann Ewald of the German Jagers participated in the fight and described the engagement.

> *The Indians as well as the Americans defended themselves like brave men against all sides where they were attacked, so that a hot fight resulted in five or six parties where the heavily wooded terrain offered cover. By seven o' clock in the evening, however, most of the enemy were killed, partly shot dead and partly cut down by the cavalry. No Indians, especially, received quarter, including their chief called Nimham and his son, save for a few.*[23]

General Scott reported that only fourteen Indians returned to camp and that Major Steward lost another twenty men.[24] Ewald noted that English losses were also high with forty dead and more wounded.[25]

---

[22] Hoth, ed., "General Washington to Brigadier General Charles Scott, 14 August, 1778," *The Papers of George Washington*, Vol. 16, 313
[23] Ewald, 145
[24] Hoth, ed., "General Scott to General Washington, 31 August, 1778," *The Papers of George Washington*, Vol. 16, 448
[25] Ewald, 145

The challenge of reconnoitering the area south of White Plains with mostly infantry proved quite demanding for General Scott and his light corps, so the arrival of Major Lee's three troops of cavalry was heartily welcomed. Lee and his dragoons, approximately 100 strong, were assigned to patrol the east side of the Bronx River.[26] They did so in numerous small detachments and patrols.

In mid-September indications that the British were planning a major movement from New York reached General Washington. Efforts to determine their destination, however, failed. While he waited for more information, Washington speculated about British intentions in a letter to his brother.

> [The British] *are busily preparing...for something – Whether to operate against our Posts in the highlands (on Hudsons River) & this army – whether for a move Eastwardly & by a junction of their Land and Naval force* [in Rhode Island] *attempt the destruction of the French Fleet at Boston, & repossession of that Town – or whether to leave us altogether for the purpose of reinforcing Canada...is a matter yet to be decided....*[27]

In an effort to guard against every possibility, General Washington retreated 40 miles northward and posted troops near Fishkill and Fredericksburg, New York and Danbury, Connecticut.[28] He also reinforced West Point, across the Hudson River. Major Lee and the rest of General Scott's light troops continued to screen the main army, withdrawing northward a short distance to Washington's old encampment at White Plains.

---

[26] Hoth, ed., "General Scott to General Washington, 9 September, 1778," *The Papers of George Washington*, Vol. 16, 547-48

[27] Chase, ed., "General Washington to John A. Washington, 23 September, 1778," *The Papers of George Washington*, Vol. 17, 111

[28] Chase, ed., "General Washington to Henry Laurens, 23 September, 1778," *The Papers of George Washington*, Vol. 17, 93

After a week of grueling patrols, made more difficult by the increased distance between the American light corps and the British lines, Scott's troops discovered a large enemy force of 3,000 men marching north.[29] It proved to be a foraging party interested more in gathering provisions than attacking the Americans.

While General Scott's light corps closely observed the British foragers, a much larger British force crossed the Hudson River and marched into New Jersey to gather their own forage. A portion of this force under General Charles Grey caught Colonel George Baylor's 3rd Continental Light Dragoons by surprise in late September near Tappan, New York and decimated the American cavalry regiment.

Accounts from the American survivors, many severely wounded by British bayonets and sabers, claimed that the British refused to grant quarter (accept surrender) to the surprised Americans. One survivor recalled:

> *When the enemy entered the barn where his troops lay, he and his men asked for quarter, and were refused; that the British captain, Bull, after inquiring how many of the rebels were dead, on being told the number, ordered all the rest to be knocked on the head, and that his orders were executed on five or six of the wounded.*[30]

Survivor after survivor gave a similar account of British cruelty. Only a handful of Baylor's dragoons escaped unharmed. Colonel Baylor was not one of them. Severely wounded in the attack, Baylor never returned to duty and died in 1784.[31]

---

[29] Chase, ed., "General Scott to General Washington, 23 September, 1778," *The Papers of George Washington*, Vol. 17, 106

[30] Thayer, 150

[31] Chase, ed., Footnote 1, *The Papers of George Washington*, Vol. 17, 166-167

The news of Baylor's defeat swept through the American army. Two days later, Major Lee and his dragoons helped inflict a small degree of vengeance upon the enemy. Lee and his cavalry joined Colonel Richard Butler on an early morning patrol that encountered a party of German jagers. Lee and Butler routed the German riflemen, inflicting nearly thirty casualties while suffering no losses of their own.[32] Colonel Butler proudly extolled the courage of his infantry and Major Lee's cavalry to General Scott.

> *I Assure you Sir I have never seen Greater bravery Display'd than on this occasion, by both foot and horse, I Cant Say too much in praise of the whole as they Not only Shew'd bravery but good order also.... I Am much indebted to the Bravery of Captn Graham & his* [infantry] *officers, Lt. Rudolph of Major Lees Corps, the whole in short (that I saw) in Action would have done honour to Any Corps in Europe....*[33]

Like Colonel Butler, General Scott was pleased by the performance of his men and sent out a steady stream of patrols to skirmish and observe the enemy's movements. Major Lee and his dragoons were a huge asset for such activity and participated in many of these patrols. Occasionally they encountered the enemy, but the British were careful not to stray too far from their newly advanced lines. By mid-October, General Clinton's troops withdrew (with full wagons of forage) to their original position at King's Bridge across from Manhattan Island.

General Scott remained above White Plains but ordered his patrols to extend their movements to the enemy's new lines. Out of support range from Scott's main body, these patrols

---

[32] Chase, ed., "General Washington to General Greene, 1 October, 1778," *The Papers of George Washington*, Vol. 17, 215
[33] Chase, ed., "Colonel Butler to General Scott, 30 September, 1778," *The Papers of George Washington*, note 1, Vol. 17, 209

faced the possibility of being cut off and destroyed. Luckily for the Americans, General Clinton adopted a very defensive posture and his troops rarely ventured beyond their lines to confront the American patrols. As a result, very little combat occurred. A frustrated Major Lee, eager to engage the enemy, complained to General Scott about the new British tactics in late October.

> *Altho I have been indefatigable in my exertions to strike a blow on some part of the enemy; I find myself baffled by the new system of conduct they have introduced. Neither officer nor soldier is permitted to advance on any occasion beyond their picquets...and the out-posts are drawn in, and established under cover of the fort. No decoy can take effect, they will not pursue. Gen. Kniphausen has ordered the advanced corps to act totally on the defensive....*[34]

Lee concluded his letter with speculation that the British were preparing to evacuate New York.

Numerous reports of frenzied British naval activity helped convince many in the American camp that General Clinton did indeed plan to evacuate New York. Washington hoped that this was true, but waited for definitive proof before he moved his army again. The proof never materialized; the activity around New York was not the preparation for an evacuation, but rather, an expedition to strike Savannah, Georgia.

By mid-November, General Washington realized that General Clinton intended to stay in New York. With the campaign season at an end, Washington disbanded the light corps and ordered the men to return to their original units.[35]

---

[34] Chase, ed., "Major Lee to General Scott, 30 October, 1778," *The Papers of George Washington*, Vol. 17, 650 note 1

[35] Edward G. Lengel, ed., "General Washington to Colonel David Henley, 27 November, 1778," *The Papers of George Washington*, Vol. 18, 309

Recounting the logistical nightmare of maintaining the army in one location for the winter (Valley Forge), Washington chose to disperse the army to a number of sites in Connecticut, New York, and New Jersey. General Washington sent his cavalry regiments even further away, ordering Bland's regiment and the remnants of Baylor's to Virginia, Moylan's regiment to Pennsylvania, and Sheldon's New England cavalry regiment to Connecticut.

Major Lee and his corps were ordered to New Jersey, first Burlington, between Trenton and Philadelphia, but when this proved to be a poor location for forage, Lee's corps moved south to Woodbury, southeast of Philadelphia.

While Lee and his men were en route to their winter quarters the young major exchanged letters with General Washington regarding horses.

> *Head Quarters, Middle Brook, NJ*
> *14th Decr. 1778*
>
> Dear Lee,
>
> *The bearer has my horse in exchange for your Mare.... I do not want the Mare to be sent to me, my wish is to send her and the other to my seat in Virginia – Do you know of any good & safe oppertunity of doing it? Were you not to have sent a Horse to me to look at for my own riding? Will you come & dine with me to day?*
>
> *I am Yr. sincere friend & Affecte Hble Servt.*
>
> *Go: Washington* [36]

Major Lee was only a few miles from headquarters and responded to the letter the same day.

---

[36] Lengel, ed., "General Washington to Major Henry Lee, Jr., 14 December, 1778," *The Papers of George Washington*, Vol. 18, 409

> *I have the honor of your Excelleny's letr accompanied with your two horses.... Altho, I omitted sending to H. Quarters yesterday the horse I mentioned some time ago, I had not forgot it, but postponed it purposely, till I could know with more exactness the abilitys of the horse.... Wishing to take the field in the spring with my corps in perfect order, I must procrastinate the honor of waiting on your Excellency at present, as I hasten to Winter-quarters to commence preparations.*
>
> *I have the honor to be unalterably your Excellencys affectionate & obedt servt.*
> *Henry Lee jr.*[37]

The fact that Major Lee declined an invitation to dine with the commander-in-chief of the American army says a lot about Lee's confidence in his relationship with Washington. Few in the army would have been so bold to decline such an invitation and opportunity.

Lee and his corps remained in Burlington for only two weeks before insufficient forage prompted him to write to head quarters for permission to seek a better winter location.

> *The very great scarcity of forage in this place obliges me to apply for change of quarters. We have never been so destitute, even in the most difficult part of the campaign; and I am fully persuaded from my own knowledge, that it is impossible to winter here without ruining our horse.*[38]

---

[37] Lengel, ed., "Major Henry Lee Jr. to General Washington, 14 December, 1778," *The Papers of George Washington*, Vol. 18, 409-410
[38] Lengel, ed., "Major Henry Lee Jr. to Tench Tilghman, 30 December, 1778," *The Papers of George Washington*, Vol. 18, 530

In late December, Lee's corps moved a days march further south to Woodbury, New Jersey and took full advantage of the plentiful supplies to rest and restore their horses and themselves.

Chapter Five

"Major Lee has performed a most gallant affair"
1779

Major Lee and the officers of his partisan corps spent much of the winter recruiting more dragoons to fill out the ranks of Lee's three troops of cavalry. The enlistments for the men in Lee's original 5th troop expired at the end of 1778 and while most of the officers (commissioned and non-commissioned) re-enlisted, nearly all of the rank of file cavalrymen of the 5th troop left Lee's corps.[1] Lee replaced many of the men with recruits from New Jersey.[2]

The winter proved to be rather uneventful for Major Lee and his corps. Lee occasionally turned up at headquarters in Middle Brook to report to General Washington or serve on a court martial tribunal, and he and his corps had the honor of leading a military procession for the French Foreign Minister, Conrad Alexander Gerard in the early spring, but most of Lee's activities in early 1779 involved the recruitment, equipping, and training of his partisan corps.[3]

Despite Lee's relative inactivity, there was still plenty of demand for his services. In May, officials from Virginia requested that General Washington transfer Lee's corps to Virginia to help quell disturbances on their western frontier. General Washington politely declined the request.

---

[1] John W. Hartmann, *The American Partisan: Henry Lee and the Struggle for Independence: 1776-1780.* (Burd Street Press, 2000), 84
[2] Ibid., 85
[3] Ibid.

> *From the present condition and arrangement of the Cavalry, I can not think that Major Lee's Corps can be sent to Virginia consistently with the general service.... It is but small, and I should hope that there would be more than a sufficiency of Militia...to restrain [the Enemy's] excursions. Major Lee's corps of Horse, from the broken and shattered condition of Moylan's and Sheldon's, is what I have principally to depend on in this line – and without it, we might experience at least, great inconveniences.*[4]

Once again, General Washington made clear his reliance on Major Lee and his corps. It was a reliance that Washington would turn to again and again over the next two years.

A week before General Washington declined the request to send Lee to Virginia, Major Lee wrote to Washington with a request of his own. Lee wanted to expand his corps by one company and he had one particular unit in mind.

> *In consequence of Captain [Allen] McCleans request, I do myself the honor to mention to your Excellency, the very great satisfaction I should receive, on having the Captain and his company annexed to the Corps under my command.*
>
> *Exclusive of the great advantage resulting from a small, choice, and alert body of infantry, I flatter myself with deriving particular assistance from the approved zeal, prudence and activity of the officer who commands them.*[5]

---

[4] Edward G. Lengel, ed., "General Washington to Virginia Delegates, 25 May, 1779," *The Papers of George Washington*, Vol. 20, 629

[5] Lengel, ed., "Major Henry Lee Jr. to General Washington, 18 May, 1779," *The Papers of George Washington*, Vol. 20, 528

Captain Allen McLane of Delaware had already distinguished himself with a superb military service record. McLane saw action at Long Island, White Plains, and Trenton and Princeton in 1776-77. His brave conduct at Princeton earned McLane a promotion to captain and he commanded troops at Brandywine and Germantown.[6] Like Lee, McLane saw important service at Valley Forge in 1778 commanding a party of observation comprised of infantry and cavalry.[7] Unlike Lee, McLane participated in the Monmouth campaign and reportedly helped capture hundreds of British stragglers.[8]

Although Major Lee's letter to Washington clearly stated that Captain McLane had requested a transfer to Lee's corps, the history of these two officers suggests otherwise. In fact, their first encounter in the winter of 1778 nearly ended in a duel between the two men. Captain McLane had approached Major Lee on behalf of a Delaware farmer whom Lee had accused of having Tory sympathies and Lee apparently did not appreciate the interference. McLane recalled years later in his memoirs that Lee insulted him, then suggested that the dispute be settled over a pair of pistols before finally threatening to arrest McLane. Lee eventually calmed down and let the matter drop.[9]

Another factor that suggests that Captain McLane did not support his transfer to Lee's corps was McLane's reaction to the news that the transfer was permanent. The only infantry that had served with Major Lee up to this point of the war had been temporary detachments assigned for a specific time period or mission. Captain McLane assumed that a similar arrangement had been made for his company but was stunned to discover otherwise a few weeks after he joined Lee's corps in mid-June.[10] A flurry of letters between McLane, Lee,

---

[6] Hartmann, 86
[7] McLane Papers, Vol. 1, Memoirs, New York Historical Society, 47
[8] Hartmann, 88
[9] McLane Papers, Vol. 1, Memoirs, New York Historical Society, 75
[10] Hartmann, 90

Washington, and Congress resulted in the formal incorporation of Captain McLane's company into Lee's corps as Lee's fourth troop.[11] Years later McLane, who had prided himself on having his own separate command up to 1779, cited his transfer to Lee's corps as the death blow to his military career.[12] Recognition from this point on for Captain McLane would only come through the reports and good graces of Major Lee.[13] Despite McLane's somber reflections late in life, opportunities to enhance his reputation still arose for Captain McLane in the summer of 1779.

In early June, General Clinton marched a large British force north from New York and seized Stony Point and Verplancks, two important American outposts just 12 miles south of West Point on the west and east banks of the Hudson River. Part of General Washington's response was to order Major Lee and his corps northward to patrol southern New York. Washington's instructions to Lee were clear.

> *The intention of your command will be to countenance the militia, plague the enemy and cover the country from the depredations of their light parties, as much as possible. The enemy have now a body at Kings ferry and appear to be establishing a post at Stoney point to which quarter your attention is principally to be directed.*
>
> *I leave you at perfect liberty to dispose of yourself as you think most proper for answering the purpose I have mentioned consistent with the security of your corps. Your utmost vigilance and attention will be necessary, as you will be entirely detached and unsupported, and will act in a very disaffected*

---

[11] Ibid., 90-91

[12] William H. Richardson, *Washington and the Enterprise Against Powles Hook*, (Jersey City, NJ: New Jersey Title Guarantee and Trust Co., 1929), 32

[13] Ibid.

> *country, the inhabitants [of which] will give the enemy every kind of intelligence, to enable them to take advantage of your situation. You will take every measure in your power to acquire information of their situations, movement and designs and give me the earliest advice of every occurrence.*
> *P.S. I wish you to exert yourself to keep up the spirits of the militia...*[14]

While Lee sought to bolster the militia, prevent enemy raids into southern New York and northern New Jersey, and gather intelligence on the enemy post at Stony Point, he also offered advice to General Washington aimed at addressing another significant problem in the army, desertion. Lee suggested a rather harsh method to deal with deserters, summary execution followed by beheading and display of the deserter's head as an example for others. General Washington rejected Lee's proposal as too harsh.

> *The measure you propose of putting deserters from our Army to immediate death would probably tend to discourage the practice. But it ought to be executed with caution and only when the fact is very clear and unequivocal. I think that the part of your proposal which respects cutting off their heads and sending them to the Light Troops had better be omitted. Examples, however severe, ought not to be attended with an appearance of inhumanity otherwise they give disgust, and may excite resentment rather than terror.*[15]

---

[14] Fitzpatrick, ed., "General Washington to Major Henry Lee Jr., 6 June, 1779," *The Writings of George Washington*, Vol. 15, 232-233
[15] Fitzpatrick, ed., "General Washington to Major Henry Lee Jr., 9 July, 1779," *The Writings of George Washington*, Vol. 15, 388

Unfortunately, Major Lee took the liberty to act on his proposal before Washington's refusal reached him. Washington was appalled to hear that Lee had already decapitated an accused deserter.

> *I fear it will have a bad effect both in the army and in the country.... You will send and have the body buried lest it fall into the enemy's hands.*[16]

Although Washington disapproved of Lee's harsh treatment of deserters, he was pleased with Lee's efforts to gather intelligence about the British post at Stony Point.

Stony Point

General Washington had determined to strike back at Stony Point soon after its fall to the British in June 1779. He assigned the task of planning and executing an attack on Stony Point to General Anthony Wayne and his 1,200 man light infantry corps. Wayne needed accurate intelligence about the post so Major Lee was tapped to, *"obtain the best knowledge,"* of the garrison possible.[17] Lee constantly patrolled near Stony Point, interviewing locals and observing the post from a distance. Near the eve of the attack Lee was even able to get a man inside the fort. Captain Allen McLane, disguised as "a lowly backwoodsman," escorted the mother of a local captive on a visit to the fort and risked apprehension as a spy.[18] McLane's intelligence helped Wayne finalize his plans. The attack was set for July 15th at midnight.

At first glance, Stony Point looked impregnable, despite the intelligence gathered by Lee and McLane. The British garrison at Stony Point numbered over 600 men who defended

---

[16] Fitzgerald, ed., "General Washington to Major Henry Lee Jr., 10 July, 1779," *The Writings of George Washington*, Vol. 15, 399

[17] Fitzgerald, ed., "General Washington to Major Henry Lee Jr., 28 June, 1779," *The Writings of George Washington*, Vol. 15, 339

[18] Hartmann, 93

a rocky promontory rising 150 feet above the river.[19] The post was protected by water on three sides and connected to land by a swampy morass that was often submerged at high tide. Two lines of earthworks with abattis, (sharp sticks that protruded toward the enemy and acted as barbed wire) were placed in front of the earthworks. Several redoubts with cannon were also erected and British ships in the Hudson River provided additional support.

General Wayne's plan called for three detachments to simultaneously attack the fort from three directions. The smallest detachment consisted of two companies of North Carolina troops under Major Hardy Murfree. They were the only troops allowed to load their muskets and were ordered to, *"keep up a perpetual; and Galling fire,"* on the center of the enemy's line only after the British pickets became alarmed by the other detachments.[20] General Wayne hoped that Major Murfree's fire would draw the enemy's attention to the center of their line and away from their flanks.

Lieutenant Colonel Richard Butler, with 300 Maryland and Pennsylvania troops, would assault the northern side of the fort while Colonel Christian Febiger and General Wayne, with the main American force of 700 men waded through Haverstraw Bay and attacked the southern side of the fort. Twenty man advance parties led both flank columns to remove the abattis and other obstructions for the troops behind them.[21] Major Lee's corps trailed the column on the march to Stony Point and was held in reserve.[22]

---

[19] Don Loprieno, *The Enterprise in Contemplation: The Midnight Assault of Stony Point*, (Westminster MD: Heritage Books, 2004), 6-7

[20] Charles Stille, "General Wayne's Order of Battle, 15 July, 1779," *Major-General Anthony Wayne and the Pennsylvania Line in the Continental Army*, (Port Washington, NY: Kenniket Press, Inc, 1968), 402
First published in 1893

[21] Ibid.

[22] Hartmann, 93

Stony Point

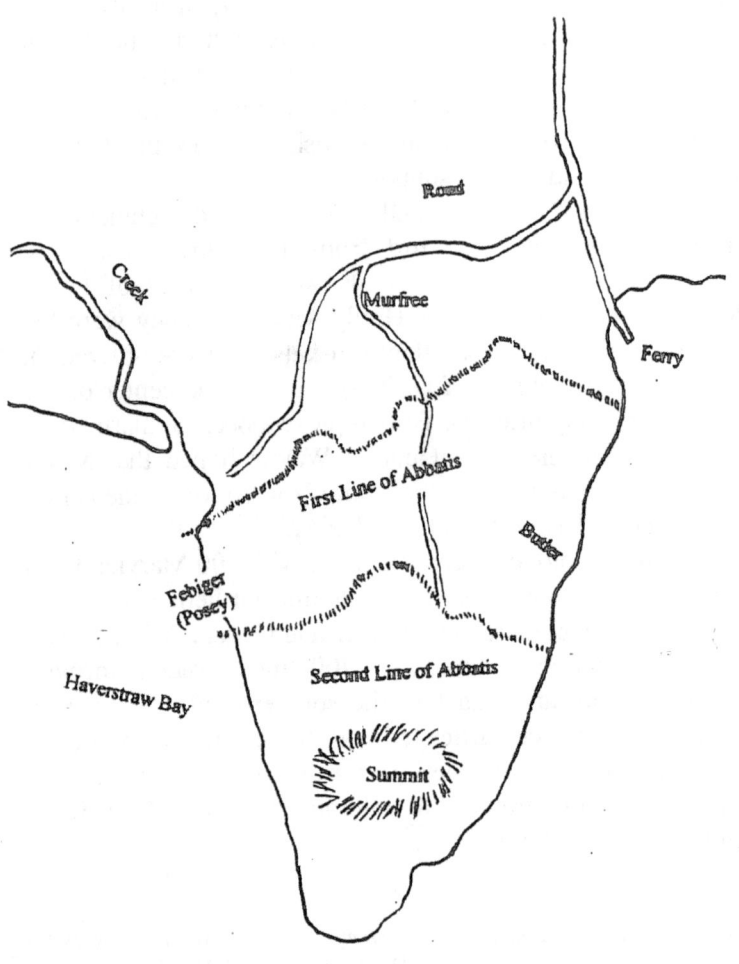

The attack began at 12:20 a.m., twenty minutes behind schedule due to high water in the bay.[23] As the Americans approached the fort, British sentries grew alarmed. Although they could not see the Americans, they heard them and fired in their direction. Lieutenant John Ross, the commander of the British picket guard, initially believed that his men were trigger happy and had fired at the wind.[24]

British sentries on the left side of the line, however, alarmed by the firing of their comrades, soon noticed movement in Haverstraw Bay. It was the main American column approaching from the south.[25] The British pickets fired and withdrew to the outer earthworks and braced for an attack. On the left of their line, a twelve pound cannon illuminated the night. Lieutenant William Horndon of the Royal Artillery recalled that,

> *By the lite occasioned by the flash of the Gun I could perceive a Body of* [the Americans] *coming thro' the Water; upon the left... I attempted to bring the Gun to bear upon them, but could not effect it, the Embrazure being too confined.*[26]

The main American column pressed on, wading through waist deep water to pass around the enemy's outer abattis. Lieutenant Colonel William Hull described the attack:

---

[23] Henry Johnson, "General Wayne to General Washington, 17 July, 1779," *The Storming of Stony Point on the Hudson, Midnight, July 15, 1779: Its Importance in the Light of Unpublished Documents*, (New York: James T. White, 1900), 209
[24] Laprieno, 25
[25] Ibid., 26
[26] Ibid., 176

The beach was more than two feet deep with water, and before the right column reached it we were fired upon by the out-guards which gave the alarm to the garrison. We were now directly under the fort, and closing in a solid column ascended the hill, which was almost perpendicular. When about half way up our course was impeded by two strong rows of abatis, which the forlorn hope had not been able entirely to remove. The column proceeded silently on, clearing away the abatis, passed to the breastwork, cut and tore away the pickets, cleared the cheveaux de fries at the sally port, mounted the parapet and entered the fort at the point of the bayonet. All this was done under a heavy fire of artillery and musketry, and as strong a resistance as could be made by the British bayonet.[27]

General Wayne, who was wounded in the attack also described the assault.

The troops [were under] *the most pointed Orders not to attempt to fire, but put their whole dependence on the Bayonet – which was faithfully & Literally Observed, --neither the deep Morass, the formidable & double rows of abbatis or the high & strong works in front & flank could damp the ardor of the troops – who in the face of a most tremendous and Incessant fire of Musketry & from Artillery loaded with shells & Grape-shot forced their way at the point of the Bayonet thro' every Obstacle, -- both Columns meeting in the Center of the Enemy's work nearly at the same Instant.*[28]

---

[27] Johnson, 191
[28] Johnson, 209

The British garrison was overwhelmed by Wayne's troops, who scored a stunning victory at the point of their bayonets. At the cost of 15 Americans killed and 83 wounded, Wayne's men had seized Stony Point and inflicted over 550 casualties on the enemy (approximately 100 killed and wounded and 450 captured).[29]

American morale soared throughout the country. In the days following the successful assault on Stony Point, Major Lee and his corps resumed their patrols and reconnaissance activities in southern New York and northern New Jersey. In late July, Lee wrote to General Washington with a bold proposal to attack another strong British outpost on the Hudson River less than two miles from New York City, Powles Hook.[30]

Powles Hook

Like Stony Point, Powles Hook sat on a peninsula that jutted into the Hudson River and although its elevation was barely above sea level, the 400 man garrison at Powles Hook was protected on three sides by the Hudson River and from the west by a large marsh that flooded at high tide.[31] In fact, the only land approach to the fort was over a long causeway through the marsh and over a drawbridge that spanned a tidal moat dug across the peninsula.[32] A ring of abattis encircled the post which included two fortified redoubts bristling with cannon and two block houses. All of this was within the shadow of New York City and the British fleet which made Powles Hook a formidable fort to attack.[33]

---

[29] Boatner, 1066
[30] Fitzpatrick, ed., "General Washington to Major Henry Lee Jr., 28 July, 1779," *The Writings of George Washington*, Vol. 15, 498
 Note: Washington refers to Lee's proposal in the letter above.
[31] Hartmann, 106-107
[32] Ibid., 107
[33] Ibid.

Powles Hook

The British Post of "Paulus Hook"
From the painting by Edward L. Henry, N. A.

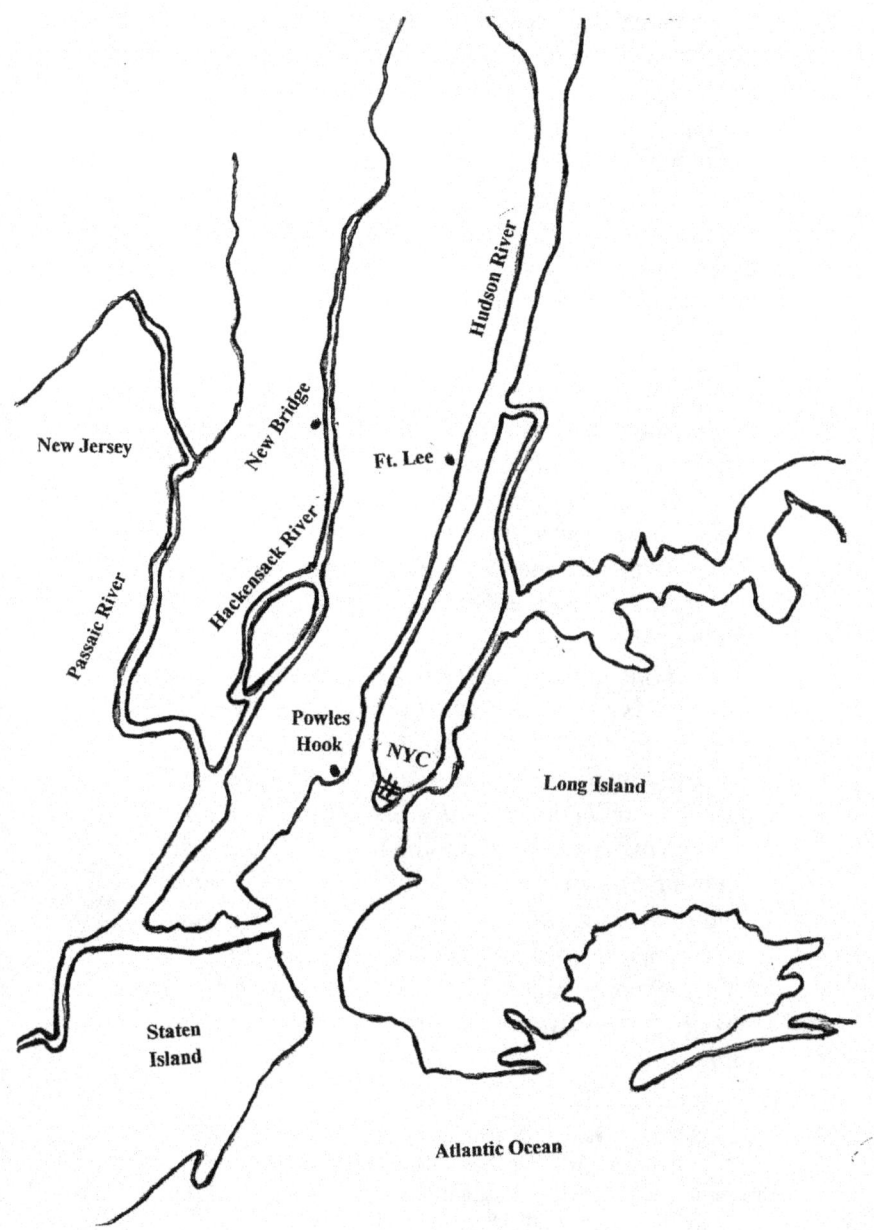

Although General Washington was initially skeptical of Major Lee's proposal to strike Powles Hook, Lee was given permission to gather more intelligence on the post in preparation for a possible attack.[34] By mid-August, Major Lee had acquired enough information on the garrison and had adjusted his plan of attack sufficiently to gain General Washington's cautious consent.

Since most of the troops that Lee proposed for the operation were to come from General Stirling's division, Washington wrote to Stirling to solicit his opinion on Lee's proposal.

> *I have had in contemplation an attempt to surprise the enemys Post at Powlus Hook and have employed Major Lee to make the necessary previous inquiries. He will inform you of what has passed between us. The number first proposed for the enterprise was 600, but these appeared to me too many to hazard for an object of inferior importance: But by the inclosed letter...Major Lee proposed to reduce the number to 400, three hundred of which to be employed in the attack. As the success must depend on surprise these appear to me sufficient to effect the purpose and as many as ought to be hazarded in the attempt.... Your Lordship will consult Major Lee fully and if, upon the whole you deem the undertaking eligible you have my consent to carry it into execution.... But I need not add that the greatest caution will be necessary not to give a suspicion of our design and to keep it a matter of profound secrecy. The least alarm, would probably occasion disappointment and ruin the party.*[35]

---

[34] Fitzgerald, ed., "General Washington to Major Henry Lee, 28 July, 1779," *The Writings of George Washington*, Vol. 15, 498

[35] Fitzgerald, ed., "General Washington to Lord Stirling, 12 August, 1779," *The Writings of George Washington*, Vol. 16, 83-84

General Stirling approved Lee's plan, and a week later, on August 18th, Major Lee marched southward from Stirling's headquarters at New Bridge (Hackensack, NJ) towards Powles Hook. With him were 350 Virginia and Maryland troops and Captain Allen McLane's troop of infantry.[36] While McLane's men were the only ones from Lee's corps assigned to the assault force, Lee's dragoons played an important role. They guarded the roads from Manhattan Island and screened Lee's detachment along the eighteen mile march route.[37]

Major Lee tried to time the long march so that his attack commenced at half past midnight, a few hours before high tide, but his guide led the column astray. As a result, Lee and his men spent crucial hours and energy unnecessarily marching through difficult terrain. Lee angrily recalled that,

> *After filing into the mountains, the timidity or treachery of the principal guide prolonged a short march into a march of three hours; by this means the troops were exceedingly harassed; and, being obliged to pass through deep, mountainous woods to regain our route, some parties of the rear were unfortunately separated.*[38]

Despite the disappearance along the march of a quarter of his detachment, Lee continued on towards Powles Hook. When they reached the edge of the marsh around 3:30 a.m., Lee sent Lieutenant Michael Rudolph of Captain McLane's troop ahead to reconnoiter the fort. Lieutenant Rudolph reported that the fort was quiet and, despite the approach of high tide, the moat was still fordable.[39] With no time to

---

[36] Frank Moore, ed., "Extract of a letter from an officer at Paramus," *Diary of the American Revolution*, Vol. 2, (Charles T. Evans: NY, 1863), 207

[37] Hartmann, 109

[38] Moore, ed., "Extract of a letter from an officer at Paramus," *Diary of the American Revolution*, Vol. 2, 207

[39] Ibid.

organize his men into three columns as originally planned, Lee ordered them to advance as they were. He explained his decision in a letter to General Washington.

> *I found my* [original plan of attack] *impracticable, both from the near approach of day, and the rising of the tide. Not a moment being to spare, I paid no attention to the punctilios of honor or rank, but ordered the troops to advance in their then disposition.*[40]

Lee pushed his men forward, determined, "*to leave my corpse within the enemy's line,*" if the attack failed.[41] Captain Levin Hardy commanded a company of Maryland troops in the assault and described the advance.

> *We had a morass to pass of upwards two miles, the greatest part of which we were obliged to pass by files, and several canals to ford up to our breast in water. We advanced with bayonets, pans open, cocks fallen, to prevent any fire from our side....*[42]

Major Lee credited two officers in particular with leading the assault.

> *The forlorn hopes, led by Lieutenant McCallister...and Lieutenant Rudolph, marched on with trailed arms in most profound silence...the first notice to the garrison was the forlorn* [troops] *plunging into the canal. A firing immediately*

---

[40] Moore, ed., "Extract of a letter from an officer at Paramus," *Diary of the American Revolution*, Vol. 2, 207
[41] William B. Reed, "Henry Lee to President Reed, 27 August, 1779," *Life and Correspondence of Joseph Reed*, Vol. 2, (Lindsay and Blakiston: Philadelphia, 1847), 126-27
[42] Reed, "Levin Handy to George Handy, 22 July, 1779," *Life and Correspondence of Joseph Reed*, Vol. 2, 126

> *commenced from the block-houses, and along the line of abattis, but did not in the least check the advance of the troops.*[43]

Lee's men rushed the outer works and stormed into the fort. A bit of fortune shined on them when they discovered that the main gate was open in expectation of the return of a large Tory patrol. This also meant that the garrison was weaker than anticipated. The Americans poured into the fort, but they still had two strong redoubts and two fortified block houses to overcome.

Lieutenant McAllister, supported by Major John Clark of Virginia, easily overwhelmed the dazed defenders of the center redoubt, capturing six cannon and the post's colors.[44] At the same time, Lieutenant Michael Rudolph, supported by Captain Robert Forsyth and Captain McLane, stormed one of the blockhouses.[45] The main barracks of the post, filled with invalids and camp followers, also fell quickly, but the fort's commander, Major Nicholas Sutherland and about 25 Hessians successfully defended the second redoubt.[46]

Up to this point the Americans had not fired their muskets, relying instead on surprise and their bayonets to subdue the enemy. With the garrison thoroughly aware of the American presence and resistance continuing in one of the redoubts, musket fire might have been useful. It was not an option for most of Lee's troops, however, because most of the men ruined their powder when they waded through the flooded marsh and moat. The Americans gathered what little powder they could from the fallen and captured enemy and tried to seize the garrison's powder magazine but failed to gain access to it.

---

[43] Moore, ed., "Extract of a letter from an officer at Paramus," *Diary of the American Revolution,* Vol. 2, 207-08
[44] Hartmann, 114
[45] Boatner, 839
[46] Hartmann, 114-115

With over 150 prisoners and approximately 50 enemy killed or wounded, all at the expense of only a handful of his own men, Lee could confidently claim success, despite the continued resistance of one of the enemy redoubts. To preserve this victory, however, Lee and his detachment, along with their prisoners, had to make good their return to the American lines.[47] With alarm guns firing across the Hudson River and the British army rousing itself in New York, it was imperative that Lee begin his return march immediately. He recalled that,

> *The appearance of daylight, my apprehension lest some accident might have befallen the boats* [that Lee and the prisoners were to march to] *the numerous difficulties of the retreat, the harassed state of the troops, and the destruction of all our ammunition by passing the canal conspired in influencing me to retire at the moment of victory.*[48]

Major Lee halted the assault on the remaining enemy redoubt and ordered his detachment to march westward with their prisoners, sparing the fort's barracks from destruction because it was occupied by women and children and sick soldiers.[49] Lee also failed to spike the enemy cannon. There was just no more time left.

To reduce the chance of being intercepted by a large enemy force from New York, Major Lee followed a different route westward for the return march, one that required the detachment to cross the nearby Hackensack River by boat. Unfortunately, while on the march to the boats, Lee learned that Captain Henry Peyton's cavalry detail, tasked with

---

[47] Reed, "Levin Handy to George Handy, 22 July, 1779," *Life and Correspondence of Joseph Reed*, Vol. 2, 126

[48] Moore, ed., "Extract of a letter from an officer at Paramus," *Diary of the American Revolution*, Vol. 2, 208

[49] Ibid., 212

guarding the boats, never received word of Lee's delay and had departed with the boats at sunrise, assuming that Lee had cancelled the attack.[50] With no way across the Hackensack River, Lee was forced to retrace his march northward along his original route and risk interception by the enemy. He recalled,

> *In this very critical situation, I lost no time in my decision, but ordered the troops to regain Bergen road.... Oppressed by every possible misfortune, at the head of troops worn down by a rapid march of thirty miles, through mountains, swamps, and deep morasses, without the least refreshment during the whole march, ammunition destroyed, encumbered with prisoners, and a retreat of fourteen miles to make good, on a route admissible of interception at several points...one [enemy] party moving in our rear and another...in all probability well advanced on our right, a retreat naturally impossible to our left, under all these distressing circumstances, my sole dependence was in the persevering gallantry of the officers, and obstinate courage of the troops.*[51]

Once again, fortune shined on Lee and his men when they encountered a detachment of fifty Virginians (likely part of Lee's missing men) with dry gunpowder.[52] Lee halted long enough to distribute a few cartridges to each man and then continued on. When they reached the vicinity of Fort Lee they were met by another large body of troops, reinforcements sent by General Stirling. Lee's men could now effectively defend themselves, which is what they did when an enemy detachment suddenly emerged on their flank. After a brief skirmish, the Americans pushed on to New Bridge and safety.

---

[50] Ibid.
[51] Ibid., 209
[52] Ibid.

Major Lee's detachment reached camp around 1:00 p.m., exhausted from nearly twenty-four hours of constant activity. They had marched almost 20 miles under difficult conditions to surprise the enemy at Powles Hook. At the loss of just a handful of men, they captured over 150 enemy troops and killed or wounded another 50.[53] They then marched another 20 miles past an alarmed enemy, burdened with enemy prisoners that they guarded with virtually empty muskets.

Praise for Lee and his expedition was extensive. General Nathanael Greene informed his wife that,

> *Major Lee has performed a most gallant affair.... This expedition is thought to be more gallant than the Stoney Point.*[54]

Fellow Virginian, General George Weedon, diplomatically noted that,

> *Wayne's and Lee's enterprises add great luster to our arms and do those Gentlemen much honor.*[55]

General Wayne was not bothered by the comparisons to his successful attack of Stony Point and graciously congratulated Major Lee.[56] Someone else who was thrilled by Lee's success was the commander-in-chief, General Washington. He expressed his gratitude to Lee and his men in the general orders.

---

[53] Reed, "Levin Handy to George Handy, 22 July, 1779," *Life and Correspondence of Joseph Reed*, Vol. 2, 126

[54] Showman, ed., "General Greene to Catherine Greene, 23 August, 1779," *The Papers of General Nathanael Greene*, Vol. 4, 333

[55] Showman, ed., "General Weedon to General Greene, 20 September, 1779," *The Papers of General Nathanael Greene*, Vol. 4, 401

[56] "General Anthony Wayne to Major Henry Lee, Jr., 24 August, 1779," Historical Society of Pennsylvania

> *The General has the pleasure to inform the army that on the night of the 18th instant, Major Lee at the head of a party composed of his own Corps, and detachments from the Virginia and Maryland lines, surprised the Garrison of Powles Hook and brought off a considerable number of Prisoners with very little loss on our side. The Enterprise was executed with a distinguished degree of Address, Activity and Bravery and does great honor to Major Lee and to all the officers and men under his command, who are requested to accept the General's warmest thanks.*[57]

Praise for Lee came also from Washington's staff. Alexander Hamilton observed to his fellow aide, John Laurens, that, "*Lee unfolds himself more and more to be an officer of great capacity.*"[58] Lieutenant Colonel Hamilton also hinted at a possible character flaw in Lee when he added, "*if he* [Lee] *had not a little spice of the Julius Caesar or Cromwell in him, he would be a very clever fellow.*"[59]

Perhaps it was this spice of Caesar and Cromwell that upset a small group of officers who withheld praise for Lee and instead, directed accusations of wrong doing at him. The first hints of disgruntlement arose from General Stirling, who seemed bothered with the degree of praise lavished on Lee instead of himself. Stirling questioned the legitimacy of placing Major Lee, a cavalry officer, in command of what was primarily an infantry detachment. General William Woodford and General Peter Muhlenberg, both fellow Virginians, chimed in with a complaint that Lee was not the senior officer

---

[57] Fitzgerald, ed., "General Orders, 22 August, 1779," *The Writings of George Washington*, Vol. 16, 149

[58] Harold Stryatt, ed., "Alexander Hamilton to John Laurens, 11 September,1779," *The Papers of Alexander Hamilton*, Vol. 2, (New York: Columbia University Press, 1962), 168

[59] Ibid.

in the detachment and that he should have relinquished command to Major John Clark.

General Washington tried to quell the objections with a long letter to Stirling. In it Washington dismissed the notion that cavalry officers had no right to command infantry detachments or that a major was unqualified to command a force the size of Lee's detachment (400 men plus Lee's cavalry).[60] Washington then outlined why Major Lee was the best choice to command the expedition.

> *This officer's situation made it most convenient to employ him to make the necessary previous inquiries. It was the best calculated to answer the purpose without giving suspicion. He executed the trust with great address intelligence and industry and made himself perfectly master of the post with all its approaches and appendages. After having taken so much pains personally, to ascertain facts and having from a series of observations and inquiries arranged in his mind every circumstance on which the undertaking must turn, no officer could be more proper for conducting it.*[61]

If General Washington thought his letter would halt the dissention, he was mistaken. Although General Stirling wisely let the matter drop, General Woodford and Muhlenberg pushed for a court martial of Lee.

The young major was stung. It especially bothered him that his own countrymen (fellow Virginians) were behind the accusations. Major Lee shared his troubles with his friend, Joseph Reed, President of Pennsylvania's Supreme Executive Council.

---

[60] Fitzgerald, ed., "General Washington to General Stirling, 28 August, 1779," *The Writings of George Washington*, Vol. 16, 190-194
[61] Ibid., 192

> *Generals and Colonels are now barking at me with open mouth. Colonel Gist, of Virginia, an Indian hunter, has formed a cabal. I mean to take the matter very serious, because a full explanation will recoil on my foes, and give new light to the enterprise.*[62]

Lee confessed that he had actually been too generous in his praise for some of the troops in the attack.

> *I did not tell the world that near one half of my countrymen (fellow Virginians) left me – that it was reported to me by Major Clarke as I was entering the marsh, -- that notwithstanding this and every other dumb sign, I pushed on to the attack.*[63]

Lee asserted that he was prepared to sacrifice his life in the attempt on Powles Hook while the efforts of many of Major Clark's Virginians were, "not the most vigorous." Lee ended the letter by assuring Reed that

> *I am determined to push Colonel Gist and party. The brave and generous throughout the whole army support me warmly..... I have received the thanks of General Washington in the most flattering terms, and the congratulations of General Greene* [and] *Wayne. Do not let any whispers affect you, my dear sir. Be assured that the more full the scrutiny, the more honour your friend will receive and the more ignominy will be the fate of my foes.*[64]

Major Lee faced eight charges at his court martial. They came down to the following accusations:

---

[62] Reed, "Henry Lee to President Reed, 27 August, 1779," *Life and Correspondence of Joseph Reed*, Vol. 2, 126
[63] Ibid.
[64] Ibid., 127

1. Major Lee withheld information from Colonel Gist.
2. Major Lee lied about the date of his commission to keep command of the detachment from Major Clarke, (who actually outranked Lee).
3. Major Lee conducted the detachment in a disorderly manner.
4. Major Lee appointed officers to detachments inappropriately.
5. Ditto
6. Major Lee ordered and conducted an unnecessary and disorderly retreat.
7. Ditto
8. Major Lee was involved in conduct unbecoming an officer.[65]

Lee's friends and allies were confident that Lee would prevail on all counts. General Nathanael Greene wrote a telling letter to General George Weedon of Virginia at the beginning of the court martial expressing his confidence in Lee's acquittal.

> *Major Lees gallant attack upon Powleys Hook I suppose you have heard of. This stroke is not of such magnitude as the Stony Point affair; but the difficulties were much greater, from the situation of the place, and the strength of the post. The attack was conducted with great spirit, and fortune favor'd it with success. The obstacles were so numerous that had not Major Lee been one of her favorite Children he must have faild. However he succeeded to the great joy of his friends. But can you believe it, he has been persecuted with a bitterness by his Countrymen, that is almost disgraceful to mention.... He has been arrested, and brought to trial, for misconduct, but there is not a shadow of evidence against him- on the*

---

[65] Fitzgerald, ed., "General Orders, 11 September, 1779," *The Writings of George Washington*, Vol. 16, 262-265

> *contrary the more the matter is enquird into, the better he appears. After passing through the furnace of affliction, he will come out, like gold seven times tried in the fire.*[66]

General Washington clearly wished the issue to be settled in Lee's favor and provided evidence in the form of a letter to discredit the charge that Lee's retreat was too hasty.

> *My principal fear, from the moment I conceived a design against the post, was on account of the difficulty of the retreat, founded on the relative situation of the post to that of the Enemy of York Island. This circumstance induced me to add...that no time should be lost in case* [the attack] *succeeded, in attempting to bring off Cannon, Stores or any other article, as a few minutes delay might expose the party...to imminent risk. I further recollect that I likewise said that no time should be spent in such case in collecting Stragglers of the Garrison, who might skulk and hide themselves, lest it should prove fatal.*[67]

Even Major Clark, (who was denied the honor of command because Lee allegedly lied about the date of his commission), helped Major Lee by testifying on his behalf. [68] After five days of testimony the court rendered its decision. Describing some of the charges against Lee as, "unsupported" and "groundless", and some of his actions as necessary and fully justified, the tribunal acquitted Major Lee with honor on all

---

[66] Showman, ed., "General Greene to General Weedon, 6 September, 1779," *The Papers of General Nathanael Greene*, Vol. 4, 364

[67] Fitzgerald, ed., "General Washington to Major Henry Lee, 1 September, 1779," *The Writings of George Washington*, Vol. 16, 217-218

[68] Hartmann, 123

eight charges.[69] Their summation of the last charge was telling of the whole trial.

> *The Court...are of opinion that Major Lee's conduct was uniform and regular, supporting his military character with magnanimity and judgment and that he by no means acted derogatory to the Gentlemen and the Soldier which characters he fills with honor to his country and the Army.*[70]

Two weeks after the verdict, Congress expressed its high regard for Major Lee's actions at Powles Hook by passing the following resolution.

> *Resolved, That the thanks of Congress be given to Major Lee, for the remarkable prudence, address, and bravery displayed by him on the* [late enterprise against Powles Hook.][71]

Congress also expressed their approval of Lee's treatment of the prisoners captured in the raid noting that he displayed, *"humanity in circumstances prompting to severity."*[72] This was a reference to the fact that Major Lee gave quarter to the enemy (accepted their surrender) during the attack, something the British neglected to do on a similar type of raid that they conducted against Colonel George Baylor's dragoons a year earlier.

Along with these resolutions of approval, Congress considered a promotion for Major Lee, but this was tabled.[73] Instead, Congress directed that a gold medal be struck and

---

[69] Fitzgerald, ed., "General Orders, 11 September, 1779," *The Writings of George Washington*, Vol. 16, 262-265
[70] Ibid.
[71] Ford, ed., "24 September, 1779," *Journals of the Continental Congress*, Vol. 15, 1099-1100
[72] Ibid.
[73] Ibid.

presented to Major Lee to commemorate his bold attack on Powles Hook.[74]

Although it took years for the medal to be designed and cast, the fact that Major Lee was awarded one of only eight such medals in the whole war must have taken some of the sting out of his court martial.

---

[74] Ibid.

Chapter Six

"Major Lee is...a man of great spirit and enterprise"

Major Lee learned of the honors bestowed upon him by Congress while on patrol in Monmouth County, New Jersey in late September. Lee and his corps had returned to action two days after his acquittal and while Lee informed his men that their mission was to gather forage for their horses, Lee was actually ordered to patrol the New Jersey coast and watch for the French navy.[1]

General Washington hoped to unite with the French and strike at New York before winter weather ended the campaign season. He expected a powerful French fleet to arrive at New York at any moment and he was eager to consult with its commander, Admiral Count D' Estaing. Major Lee was to personally deliver a letter from General Washington to Count D' Estaing outlining Washington's plan to attack New York. While Lee waited for the French fleet to arrive, he was to observe British naval activities around New York.

Determined to keep his plans a secret, Washington instructed Lee to,

---

[1] Fitzgerald, ed., "General Washington to Major Henry Lee, 13 September, 1779," *The Writings of George Washington*, Vol. 17, 279

> [Keep the mission] *a profound secret even from your officers, making your move under the colour of going to a better forage Country, and your look Outs upon the Coast may be said to be for your security from a surprise.*[2]

Major Lee centered his operations in Englishtown, New Jersey, twenty miles from the coast.[3] Constant patrols were sent throughout the countryside to forage and scout the New Jersey shore. Days and then weeks passed uneventfully and by November it was apparent that the French navy was not coming. Events in the South had conspired to disrupt Admiral D' Estaing's timetable.

In the fall of 1779, while General Washington impatiently waited for the French navy to arrive off of New York, French Admiral D' Estaing's powerful naval force sailed to Savannah, Georgia, instead. D' Estaing's 4,000 infantry troops united with American troops under General Benjamin Lincoln in an unsuccessful attempt to retake Savannah, which had fallen to the British the previous year.[4] What began as an ineffective allied siege turned into a costly allied assault of the city that was repulsed with heavy French and American casualties. After the allied defeat at Savannah, there was little hope of a joint operation in New York in 1779.

General Washington would have to wait yet another year for a chance to drive the British from New York. In the meantime, he had to lead his army through another winter, and the winter of 1779-80 was one of the harshest on record.

---

[2] Ibid.
[3] Hartmann, 151
[4] Boatner, 982

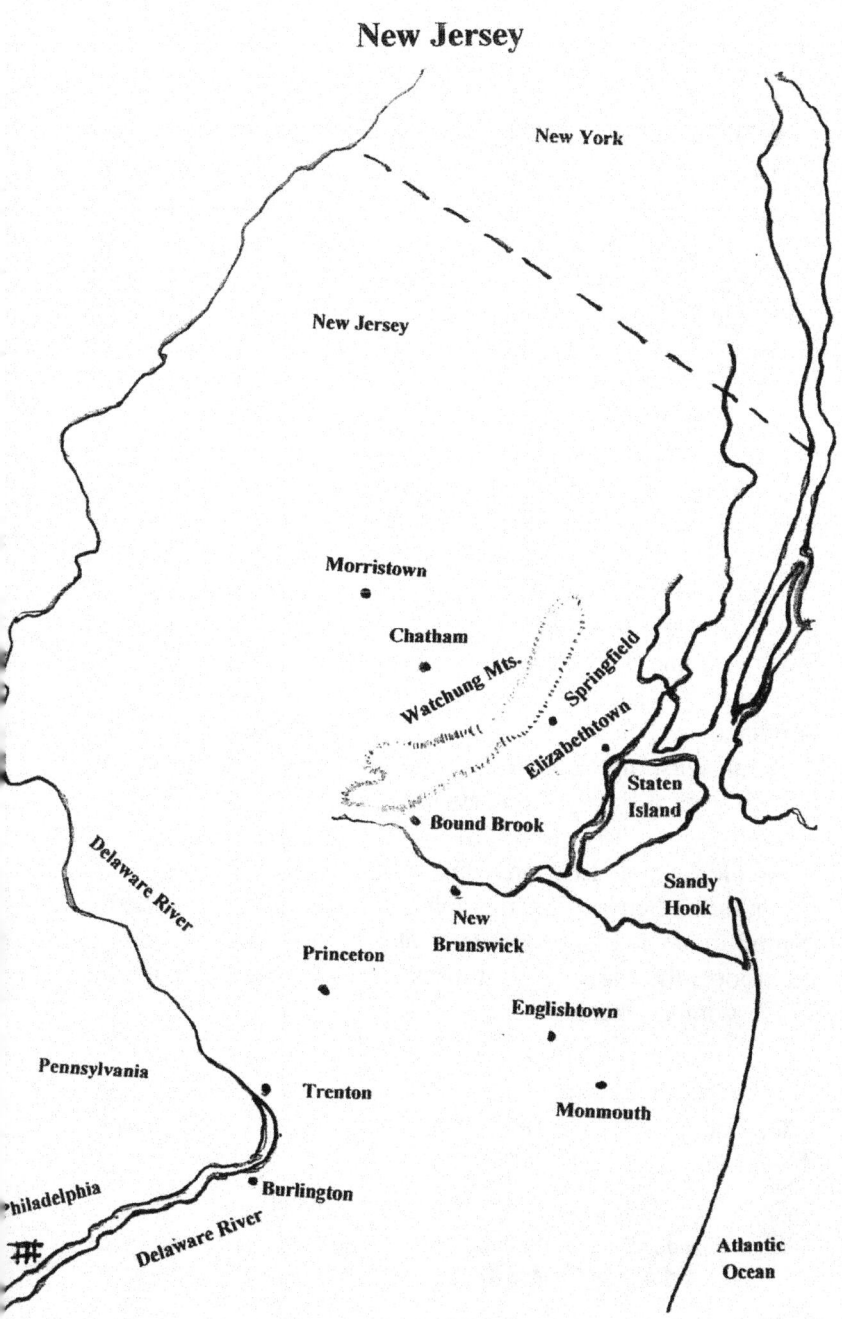

Morristown

With the approach of winter, General Washington deployed his troops in an arc above New York, extending from central Connecticut in the east through West Point and the New York Highlands to Morristown, New Jersey in the west.[5] Washington ordered most of his cavalry to winter in Connecticut near General Enoch Poor's New Hampshire brigade.[6] The Massachusetts brigades were assigned to West Point and the remainder of the army was posted at Morristown (minus the Virginia and Carolina troops who were ordered south in December to defend Charlestown, South Carolina).[7]

Although the largest part of Major Lee's partisan corps were Virginians, Lee was not ordered south. He was ordered instead, to Burlington, New Jersey where new stables awaited. In mid-December, General Washington, concerned about the lack of forage for Lee's horses at Burlington, rescinded his order and instructed Lee to remain in Monmouth County.[8] Lee was to halt the frequent contact between the local populace and enemy parties that came ashore to trade. General Washington also gave Lee permission to harass the British post at Sandy Hook (which jutted into the Lower Bay below New York Harbor) if the opportunity presented itself.[9]

While Major Lee and his corps began the winter in Monmouth, General Washington and the main army settled in behind the Watchung Mountains at Morristown. For some of the troops, their arrival in Morristown marked a return to their 1777 winter encampment. That encampment had been

---

[5] Fitzgerald, ed., "Order of Troop Cantonment, 2 December, 1779," *The Writings of George Washington*, Vol. 17, 209-211
[6] Ibid.
[7] Ibid.
[8] Fitzgerald, ed., "General Washington to Major Henry Lee, 20 December, 1779," *The Writings of George Washington*, Vol. 17, 289-290
[9] Ibid.

tolerable in part because of the mild winter. Such was not the case for the winter of 1779-80.

From the time of their arrival in early December, the American troops at Morristown struggled with the weather. Heavy snow and bitter cold swept the region while the men labored to construct huts.[10] The transport of supplies slowed to a crawl due to the heavy snow and provisions ran low for the troops. General Washington appealed to both Congress and the states in mid-December for help.

> *The situation of the army with respect to supplies is beyond description alarming. It has been five or six weeks past on half allowance, and we have not more than three days bread at a third allowance on hand, nor anywhere within reach. When this is exhausted, we must depend on the precarious gleanings of the neighbouring country. Our magazines are absolutely empty everywhere, and our commissaries entirely destitute of money or credit to replenish them. We have never experienced a like experience at any period of the war.... Unless some extraordinary and immediate exertions are made by the States...there is every appearance that the army will infallibly disband in a fortnight.*[11]

General Nathanael Greene described a similar situation to General George Weedon (who had left the army and returned to Virginia) in a Christmas day letter.

---

[10] Showman, ed., "Moore Furman to General Greene, 2 December, 1779," *The Papers of General Nathanael Greene*, Vol. 5, Note 2, 139

[11] Fitzgerald, ed., "Circular to Governors of the Middle States, 16 Dec., 1779," *The Writings of George Washington*, Vol. 17, 273-274

> *The Army is in great distress for want of Provision and forage; owing to the great Departments of the Army being kept in a starvd condition for three or four Months past for want of money. Our affairs are in a disagreeable train from the wretched state of our business of finance. This has been so badly conducted...that a thick cloud hangs over our heads at this hour threatening us with destruction.*[12]

General Greene, who commanded the quartermaster department during the crisis, attributed the primary cause for the supply shortage on a fiscal crisis that crippled the army's ability to purchase supplies. The depreciation of continental money had reached a point in which no one was willing to accept it as payment. This reluctance to trade, combined with the horrible weather conditions that continued to make travel nearly impossible, meant that necessary supplies were simply not reaching the troops.

General Greene described an even worse situation in early January, 1780.

> *Our Army is without Meat or Bread; and have been for two or three days past. Poor Fellows! They exhibit a picture truly distressing. More than half naked, and above two thirds starved. A Country, once overflowing with plenty, and now suffering an Army employed for the defense of every thing that is dear and valuable, to perish for want of food.*[13]

Corporal Joseph Plum Martin of Connecticut recalled years later in his memoir that,

---

[12] Showman, ed., "General Greene to General George Weedon, 25 Dec., 1779," *The Papers of General Nathanael Greene*, Vol. 5, 209

[13] Showman, ed., "General Greene to Moore Furman, 4 January, 1780," *The Papers of General Nathanael Greene*, Vol. 5, 230

> *We were absolutely, literally starved. I do solemnly declare that I did not put a single morsel of victuals into my mouth for four days and as many nights, except a little birch bark which I gnawed off a stick of wood, if that can be called victuals. I saw several of the men roast their old shoes and eat them, and I was afterwards informed by one of the officers' waiters, that some of the officers killed and ate a favorite little dog that belonged to one of them.*[14]

Dr. James Thacher was also at Morristown and described the condition of many of the men as, "*so enfeebled from hunger and cold, as to be almost unable to perform their military duty, or labor in constructing their huts.*"[15]

General Greene speculated that the heavy snowfall had actually helped keep the army intact by making it impossible for the troops to desert or leave in search of food in the countryside.

> *Here we are surrounded with Snow banks, and it is well we are, for if it was good traveling, I believe the Soldiers would take their packs and march, they having been without provision two or three days. The distress of the Army is very great, and not less on account of clothing, than provisions, hundreds and hundreds being without shirts, and many other necessary articles of clothing.*[16]

General Greene, like nearly everyone along the east coast, was astonished at the bitterness of the winter weather.

---

[14] Joseph Plum Martin, *Private Yankee Doolittle*, 172

[15] James Thacher, *Military Journal of the American Revolution: 1775-1783*, (NY: Corner House Historical Publishing, 1998), 185

[16] Showman, ed., "General Greene to Colonel Jeremiah Wadsworth, 5 January, 1780," *The Papers of Nathanael Greene*, Vol. 5, 236

> Such weather as we have had, never did I feel. For six or eight days it has been so extremely cold, that there was no living [outside]; the snow is also very deep, and much drifted, it is so much so, that we drive over the tops of the fences. In the midst of snow and surrounded on every side by its banks, the army has been cut off from its magazines, and been obliged to fast for several days together. We have been alternatively out of meat and bread for eight or nine days past, and without either for three or four. The distress of the army has been exceedingly great from the weather, want of clothing and provisions.[17]

Along with the weather, General Greene was amazed at the conduct of the troops who endured the hardship, "*with great patience and fortitude. They have displayed a degree of magnanimity under their sufferings which does them the highest honour....*"[18]

General Greene, like General Washington, realized however, that there was a limit to the army's patience. In desperation, Washington sent a circular letter to the magistrates of New Jersey on January 8th requisitioning provisions from the various counties.

> The present situation of the Army with respect to provisions is the most distressing of any we have experienced since the beginning of the War. For a Fortnight past the Troops, both Officers and Men, have been almost perishing for want. They have been alternately without Bread or Meat, the whole time, with a very scanty allowance of either and frequently destitute of both. They have borne their sufferings

---

[17] Showman, ed., "General Greene to an Unidentified Person, 11 January, 1780," *The Papers of Nathanael Greene*, Vol. 5, 252-253
[18] Ibid.

> *with a patience that merits the approbation and ought to excite the sympathy of their Countrymen.*[19]

Washington noted that some of his men had stolen from local inhabitants and that he feared this would increase if relief for the army did not arrive immediately. He then called upon each county in New Jersey to supply the army with cattle and grain and issued a blunt warning to local magistrates that if they did not cooperate, the army would simply impress the needed supplies.[20]

> *While I have intire confidence that you will do everything in your power to give efficacy to this requisition...I think it my duty to inform you, that should we be disappointed in our hopes, the extremity of the case will compel us to have recourse to a different mode, which will be disagreeable to me...and less convenient to the Inhabitants than the one now recommended.*[21]

On the same day he wrote to the magistrates, Washington sent instructions to a number of officers regarding the collection of provisions. Washington explained that he preferred to let the local magistrates obtain the provisions, "*but in case the requisitions should not be complied with, we must then raise the supplies ourselves in the best manner we can.*"[22] General Washington was determined to feed his troops, even if it meant seizing cattle and grain from a reluctant populace. Detachments were sent to all of the counties in New Jersey. Major Lee and his corps were assigned the three most southern counties in the state. They

---

[19] Fitzgerald, ed., "To the Magistrates of New Jersey, 8 January, 1780," *The Writings of George Washington*, Vol. 17, 362-363
[20] Ibid.
[21] Ibid.
[22] Fitzgerald, ed., "Instructions to Officers to Collect Provisions, 8 January, 1780," *The Writings of George Washington*, Vol. 17, 360-361

were expected to gather 350 head of cattle and 1500 bushels of grain in five days.[23] Although Major Lee fell short of the lofty goals set for him, he reported on January 17th that all three counties had complied as best they could and that 170 good head of cattle as well as flour and cornmeal would soon be joining the army at Morristown.[24] Many of the other American foraging parties had even greater success, and by the end of January, the supply crisis at Morristown had passed.

Prior to his foraging endeavors, Lee had applied for furloughs for many of his men and a short leave of absence for himself.[25] Both were granted, and Major Lee returned to Virginia for a few weeks in late January. There is little record of Lee's activities in Virginia, but it is reasonable to assume that the twenty-four year old returned home to Leesylvania for at least some of his leave.

While Major Lee was in Virginia, the Continental Congress approved a proposal by General Steuben to increase Major Lee's corps by 70 infantrymen.[26] The addition of the dismounted men gave Lee three troops of cavalry and three troops of infantry, but first the new men had to be recruited. Captain Allen McLane took the lead on recruiting and in just a few weeks listed 91 infantrymen on a troop strength report for Lee's corps.[27] Upon his return from Virginia in early April, Major Lee stopped in Philadelphia and submitted a request to the Board of War.

---

[23] Fitzgerald, ed., "To the Magistrates of New Jersey, 8 January, 1780," *The Writings of George Washington*, Vol. 17, 364

[24] "Major Henry Lee to General Washington, 17 January, 1780," *George Washington Papers at the Library of Congress, 1741-1799*: Series 4 (Online)

[25] "Major Henry Lee to General Washington, 13 January, 1780," *George Washington Papers at the Library of Congress, 1741-1799*: Series 4 (Online)

[26] Ford, ed., "12 February, 1780," *Journals of the Continental Congress*, Vol. 16, 159

[27] Hartmann, 166

> *The Partisan Legion consists of one Battalion of Cavalry and one Battalion of Infantry. The two Battalions form upwards of three Hundred Effective men.* [This] *number of men according to the usage of all Armies demand the Aid of three field officers.*[28]

Lee desired promotions for himself and Captains McLane and Peyton. He pointed to the arrangement of field officers for the light infantry corps as justification for increasing the number of field officers in his corps.[29] He also argued that because his officers served in an independent command, they did not have the same opportunities for promotion that officers in the continental regiments had. Officers with less service in the army had risen past Captain McLane and Captain Peyton simply because vacancies appeared in their particular state lines. But Major Lee's corps was not attached to any one state, so the opportunity for promotion in Lee's corps occurred only when a vacancy occurred within the corps itself.

Lee had approached General Washington about a promotion for Captain McLane once before, but to no avail. The recent addition of two new troops of infantry to his corps gave Lee a new opportunity to argue his case. Congress sought General Washington's view on the matter and was advised that the establishment of three field officers for Lee's corps would only create disputes in rank among other officers in the army.[30] As a result, Congress tabled Lee's request.

Although the issue of rank was important to Major Lee, he was confronted by a much larger issue when he returned to the army. Lee's corps was ordered to march to South Carolina

---

[28] "Henry Lee Jr. to the Continental Congress Board of War, 3 April, 1780," *George Washington Papers at the Library of Congress, 1741-1799: Series 4* (Online)

[29] Ibid.

[30] Fitzgerald, ed., "General Washington to Continental Congress War Board, 9 April, 1780," *The Writings of George Washington*, Vol. 18, 234-236.

on March 30th to reinforce the besieged American army in Charlestown.[31] Lee's infantry, under Captain McLane, marched to Head of Elk, Maryland in April but Lee's cavalry needed more time to prepare for the march.[32] Their horses were in bad condition after the brutal winter and needed more time to recover.[33] By the time they were able to get underway in late May, news of the fall of Charlestown arrived. Lee's dragoons were ordered to halt outside of Philadelphia while his infantry under Captain McLane waited in Portsmouth, Virginia for further orders.[34] Two weeks of inactivity ended when Lee's corps was ordered back to New Jersey in response to an enemy thrust towards Morristown.

Springfield

On June 7th, 1780 General Wilhelm Knyphausen, the commander of British forces in New York during General Henry Clinton's absence in South Carolina, led 6,000 troops into New Jersey to seize an important gap in the Watchung Mountains.[35] Knyphausen hoped to force General Washington to abandon his encampment at Morristown and then capture much of the American artillery and military stores left behind in Washington's retreat.[36] American resistance to his march, however, was stronger than Knyphasen expected. He had been led to believe by local Tories that the Americans were demoralized by the events in South Carolina, but their firm resistance suggested otherwise

---

[31] Fitzgerald, ed., "General Washington to the Commanding Officer of Lee's Corps, 30 March, 1780," *The Writings of George Washington*, Vol. 18, 183-184
[32] Hartmann, 167
[33] Ibid.
[34] Fitzgerald, ed., "General Washington to Major Henry Lee, 20 May, 1780," *The Writings of George Washington*, Vol. 18, 397
[35] Showman, ed., "General Nathanael Greene to Catharine Greene, 9 June, 1780," *The Papers of Nathanael Greene*, Note 1, Vol. 5, 10
[36] Ibid.

and prevented Knyphausen from gaining the pass before Washington reinforced it.[37]

With his objective blocked and his force outnumbered by Washington's army, Knyphausen decided to withdraw to Elizabethtown, New Jersey. He left behind the smoldering ruins of the village of Connecticut Farms (outside of Springfield) and the body of Mrs. Hannah Caldwell, the wife of a patriot leader who had been shot and buried by Knyphasusen's men.[38]

News of these events inflamed the local militia and they turned out in large numbers, but the British position at Elizabethtown was strong so the Americans did not attack. The situation turned into a standoff. General Washington was puzzled by Knyphasen's actions, first a bold advance, then a timid retreat. Lacking enough horsemen to properly reconnoiter the enemy, General Washington requested the Board of War to order Major Lee's corps to return to the main army.

> *If Major Lee's Corps is still at Philadelphia or within its vicinity, or has not advanced more than three or four days march towards the Southward, I request that you will order it to join this Army as soon as it can be done. His Horse in particular is infinitely wanted at this time. I inclose a Letter for him on this subject. The Enemy are out in force in Jersey and lie just below Springfield. They have a considerable body of Horse which we want Horse to counteract, and we want them besides for the purpose of reconnoitering &c.*[39]

---

[37] Ibid.
[38] Thacher, 198-199.
[39] Fitzgerald, ed., "General Washington to the Board of War, 8 June, 1780," *The Writings of George Washington*, Vol. 18, 488-489

Major Lee received the order to return to the main army on June 9th and he pushed his dragoons hard to do so. Lee informed Washington when they reached the outskirts of Springfield on June 11th, and assured him that despite their long ride and exhausted condition, their zeal to fight was strong.[40] This zeal appeared evident to at least one observer who commented on their martial appearance when they rode into the American camp.

> *Major Lee, from Virginia, has just arrived in camp, with a beautiful corps of light-horse, the men in complete uniform, and the horses very elegant and finely disciplined. Major Lee is said to be a man of great spirit and enterprise, and much important service is expected from him.*[41]

Although Major Lee and his corps engaged in numerous patrols and repeatedly encountered the enemy, a week passed before they got a real opportunity to display their great spirit and enterprise.

On June 19th, General Henry Clinton and 4,500 British troops from Clinton's successful South Carolina expedition arrived in New York.[42] General Clinton was not pleased by the recent activities of General Knyphausen and tried to salvage the situation by planning a pincer movement against the Americans. The timing of this operation soon broke down and it turned into a two column attack upon Springfield by Knyphausen's 6,000 man force.

The day before the attack General Washington marched the main body of his army north towards West Point. General

---

[40] "Major Henry Lee to General Washington, 11 June, 1780," *George Washington Papers at the Library of Congress, 1741-1799*: Series 4 (Online)

[41] Thacher, 200

[42] Showman, ed., "General Nathanael Greene to Catharine Greene, 9 June, 1780," *The Papers of Nathanael Greene*, Vol. 5, Note 1, 10

Clinton's activity on the Hudson River had convinced Washington that Clinton intended to strike West Point so he moved the army north as a precaution. General Nathanael Greene remained at Springfield with 1,000 continentals and an undetermined number of militia to guard against a possible enemy return there.[43] Major Lee's dragoons were with General Greene. They were posted on Greene's left flank, guarding a bridge over the Rahway River and a road that led to the rear of Springfield (and General Greene's force). Lee's dragoons also continued to regularly patrol the enemy lines at Elizabethtown.

Early in the morning of June 23rd, one of Lee's patrols discovered the enemy on the move. Major Lee reported to General Greene that a large enemy force was marching towards Springfield.[44] They marched in two columns along parallel roads a mile apart. Knyphausen's total force consisted of over 5,000 infantry, a large body of cavalry, and fifteen to twenty cannon.[45] General Greene described the dire situation he faced to Washington after the battle.

> *Our troops were so extended to guard the different roads leading to the several passes over the mountain, that I had scarcely time to collect them at Springfield and make the necessary dispositions before the Enemy appeared before the Town.... My force was small and from the direction of the roads my situation was critical. I disposed of the troops in the best manner I could, to guard our flanks, secure a retreat, and oppose the advance of their columns.*[46]

---

[43] Ibid.
[44] Showman, ed., "General Nathanael Greene to General Washington, 23 June, 1780," *The Papers of Nathanael Greene*, Vol. 5, 32
[45] Showman, ed., "General Nathanael Greene to General Washington, 24 June, 1780," *The Papers of Nathanael Greene*, Vol. 5, 34-35
[46] Ibid., 35

Troops under Colonel Israel Angell of Rhode Island held the bridge in front of Springfield and seemingly halted the enemy advance with their lone cannon and small arms fire. General Greene soon concluded, however, that Knyphausen's real objective was the bridge that Major Lee's corps was guarding on the left flank.[47]

Lee was supported by Colonel Matthias Ogden's regiment of New Jersey continentals. General Greene recalled that,

> *While the Enemy were making demonstrations to their left, their right column advanced on Major Lee. The bridge was disputed with great obstinacy, and the Enemy must have received very considerable injury, but by fording the river and gaining the point of the hill they obliged the Major with his party to give up the pass.*[48]

British Lieutenant Colonel John Simcoe of the Queen's Rangers led the assault upon Lee's position. He recalled that Lee's force was posted on the heights overlooking the bridge in such a way as to concentrate their fire on the bridge and road.[49] Using a deep gully to shield their approach, Simcoe extended his line beyond Lee's left flank and pried Lee from his position.[50]

While Major Lee defended but eventually yielded the bridge on the American left flank, Greene's main body engaged in a heated, forty minute fight for control of the bridge at Springfield.[51] Knyphausen's overwhelming numbers soon forced General Greene to withdraw to the hills northwest of Springfield. Greene's new position was strong and he

---

[47] Ibid.
[48] Ibid.
[49] Simcoe, 145
[50] Ibid.
[51] Showman, ed., "General Nathanael Greene to General Washington, 24 June, 1780," *The Papers of Nathanael Greene*, Vol. 6, 35

actually hoped that Knyphausen would pursue further so that his well protected troops might inflict more casualties on the attackers, but the German commander was content to burn the village of Springfield to the ground and retreat back to Elizabethtown.[52] General Greene sent detachments in pursuit, including Major Lee's dragoons who engaged Knyphausen's rear guard, but little damage was inflicted on either side.[53]

The Battle of Springfield marked the last significant battle in the north. Each side suffered less than a hundred casualties.[54] Besides these losses and the destruction of the village of Springfield, little was achieved by the operation.

The remainder of 1780 was marked largely by inactivity. Major Lee and his corps, bolstered by the return of Captain McLane and the infantry in August, engaged in a number of uneventful patrols and forage details, but saw little action. General Washington waited for General Clinton to act and the French to arrive, and General Clinton waited for events in the South to develop and impact the war. Major Lee was thus stuck performing routine duties.

One not so routine issue that Major Lee had to address over the summer was the disgruntlement of Captain Allen McLane. McLane's deep frustration at not being promoted, something he blamed Major Lee for, caused him to submit his resignation from the army in late August. Major Lee reported McLane's dissatisfaction to General Washington, but the commander-in-chief refused to budge on a promotion for McLane and the disgruntled captain withdrew his resignation.[55]

---

[52] Ibid.

[53] Ibid.

[54] Showman, ed., "General Nathanael Greene to General Washington, 24 June, 1780," *The Papers of Nathanael Greene*, Vol. 6, Note 1, 39

[55] Fitzgerald, ed., "General Washington to Major Henry Lee, 2 September, 1780," *The Writings of George Washington*, Vol. 19, 486-487

Note: McLane's memoirs and papers are full of bitterness and resentment towards Lee. He claimed that Lee maliciously obstructed his promotion and arranged to remove him from the legion in February 1781 as a supernumerary (extra) officer.

Captain McLane was by no means the only American officer who was frustrated by his lack of advancement in the army. Over the course of the war, hundreds of American officers resigned in protest over disputes about rank. One American officer in particular went further, however. Universally esteemed as one of America's greatest field commanders, a bitter and resentful Benedict Arnold, the hero of Quebec and Saratoga, schemed to help the British capture a vital American post on the Hudson River, West Point. Although Arnold's plot failed, he managed to escape to the British and General Clinton honored his arrangement with Arnold by making him a Brigadier General. Arnold's betrayal in late September 1780 shocked and demoralized the American army and country, but within weeks, General Washington was plotting with Major Lee to capture Arnold.

Lee's plan called for his sergeant-major, John Champe of Virginia, *"a very promising youth of uncommon taciturnity and invincible perseverance,"* to desert to the enemy and enlist under Arnold's command in the newly formed American Legion.[56] Sergeant Champe would strive to, *"insinuate himself into some menial or military birth about* [Arnold]," and when an opportunity presented itself, Champe would seize

---

McLane was in Maryland at the time, purchasing horses for the legion per Lee's instructions. Captain McLane also asserts in his papers that in March 1781, just a couple of weeks after he was forced out of Lee's Legion, he was assigned to a party of observation at the entrance of the James River by General Steuben . Two months later, after warning General LaFayette about British naval activity in the Chesapeake Bay, McLane served aboard the American ship *Congress* as a captain of Marines. He sailed to the West Indies, had an interview with Count de Grasse, admiral of the French navy, and participated in a heated naval engagement off of Charlestown, South Carolina. McLane eventually returned to Virginia, just days after the victory at Yorktown where General Washington assisted McLane in his efforts to rejoin the army as the captain of an infantry company in Colonel Charles Armand-Tuffin's legion. While all of this occurred, the company of men who served with McLane in Lee's Legion distinguished themselves with the rest of the legion in the Carolinas.

[56] Fitzgerald, ed., "General Washington to Major Henry Lee, 20 October, 1780," note 34, *The Writings of George Washington*, Vol. 20, 223

Arnold, gag him, and bring him to New Jersey for trial.[57] General Washington was particularly insistent on this last point; Arnold would stand trial for his treason. Under no circumstances was he to be killed.[58]

The plan was incredibly dangerous for Champe who, if caught, would be executed as a spy. It was concern for his reputation rather than his life, however, that caused Champe to decline the mission at first. He simply could not bear the, *"ignominy of desertion."*[59] Major Lee urged the sergeant to reconsider and accept the mission for the honor of the legion. Lee recounted in his memoirs years later that,

> [He] *entreated the sergeant to ask himself what must be the reflections of his comrades, if a soldier from some other corps should execute the attempt, when they should be told that the glory...might have been enjoyed by the Legion, had not Sergeant Champe shrunk* [from the mission].[60]

Lee noted that Champe's, *"esprit de corps could not be resisted,"* and he accepted the assignment.[61]

After a hazardous escape to the enemy, Sergeant Champe was brought before General Henry Clinton for a long interrogation during which Clinton suggested that Champe join Arnold's new unit. Champe declined and was allowed to leave, but a few days later, he enlisted in Arnold's legion as a sergeant. In that capacity, Champe hoped to gain access to Arnold's quarters and seize him one evening. All seemed to progress smoothly when all of a sudden Arnold's men were ordered to board British transport ships for a journey to

---

[57] Ibid.
[58] Fitzgerald, ed., "General Washington to Major Henry Lee, 20 October, 1780," *The Writings of George Washington*, Vol. 20, 223
[59] Lee, 398
[60] Ibid.
[61] Ibid.

Virginia. Champe's opportunity to seize Arnold and whisk him to New Jersey had passed.[62]

Sergeant Champe now found himself in the awkward position of serving under Benedict Arnold in Virginia. Six months passed before Champe was able to escape and make his way to Lee's Legion, then in South Carolina. At last the truth was revealed to his comrades and Sergeant Champe was warmly received by Lee and his men. Rewarded by both General Greene and General Washington, Champe was discharged from the army to prevent the possibility of his capture by the British.[63]

Like Sergeant Champe, the fate of Lee's entire Legion was jeopardized in the fall of 1780. Ironically, the threat came not from the enemy, but from Congress. In October 1780, while Champe was still in New York, Congress considered abolishing all of the independent corps of the army. The Board of War specifically proposed that,

> *All the separate light corps of the army, both horse and foot...be reduced on the 1st day of January next...*[and the men] *be incorporated with the troops of their respective states.*[64]

This new arrangement would have eliminated all of the army's independent units, including Lee's corps, and returned the men to regiments from their home state. This would have made it easier for each state to meet its troop quota and would have eliminated the frequent disputes between Congress and the states over who was responsible for maintaining the various independent corps of the army. Since the states were responsible for maintaining their own troops in the field, this new arrangement would simplify things.

---

[62] Ibid., 399-410
[63] Ibid., 410
[64] Ford, ed., "3 October, 1780," *Journals of the Continental Congress*, Vol. 18, 893

General Washington replied to the proposal on October 11th and confessed his dislike for independent corps, but he also asserted that he found partisan corps like Major Lee's Legion very useful.[65] Washington then made a very strong pitch to Congress in support of Major Lee.

Major Lee has rendered such distinguished Services and possesses so many Talents for commanding a Corps of this nature, he deserved so much credit for the perfection in which he has kept his Corps, as well as for the handsome exploits he has performed, that it would be a loss to the Service and a discouragement to merit to reduce him. And I do not see how he can be introduced into one of the [Virginia] *Regiments in a manner satisfactory to himself.*[66]

General Washington was not alone in his high regard for Major Lee. General LaFayette, who was himself a favorite of Washington, lavished praise on Major Lee in a letter to the commander-in-chief in late October. *"The difference made by* [Lee's] *presence in my security,"* reported LaFayette, *"is equal to the doubling of picquets and patroles."*[67] LaFayette asserted that Major Lee deserved a promotion, *"because the poor fellow cannot help having superior abilities to his Rank."*[68] LaFayette's praise for Lee did not stop with Washington, the young French general declared to the French ambassador to America, Chevalier de la Luzerne, that Major

---

[65] Fitzpatrick, ed., "General Washington to Congress, 11 October, 1780," *The Writings of George Washington*, Vol. 20, 163-164

[66] Ibid.

[67] Stanley J. Idzerda, ed., "General LaFayette to General Washington, 27 October, 1780, Note 6, *Lafayette in the Age of the American Revolution: Selected Letters and Papers*, Vol. 3, (Cornell University Press, 1980), 209

[68] Ibid.

Lee was, "*beyond compare the best officer of light infantry in the English, Hessian, or American armies on this continent.*"[69]

Such praise for Lee, particularly General Washington's, had a powerful impact on Congress. On October 21st, a new plan to reorganize the army was revealed and included the following arrangement.

> *That there be two partisan corps, consisting of three troops of mounted and three of dismounted dragoons, of fifty each, one to be commanded by Colonel Armand, and the other by Major Lee....*[70]

The next day, General Washington informed Lee to expect orders from Congress to march south.[71] Lee probably welcomed the news; it was a chance to escape from the fruitless patrols and constant forages of the summer and fall. Lee also desired to join General Nathanael Greene, his friend, who had been ordered south earlier in the month to take command of the southern army. The situation in South Carolina had deteriorated significantly since the fall of Charlestown in May. General Horatio Gates, the hero of Saratoga, was sent south in the summer to rebuild the southern army, but he was soundly defeated at the Battle of Camden in mid-August. The dispirited American army desperately needed help, and Washington hoped that General Greene was the man to provide it. On October 31st, Congress decided that General Greene could use more help.

---

[69] Ibid.
[70] Ford, ed., "21 October, 1780," *Journals of the Continental Congress*, Vol. 18, 960
[71] Fitzpatrick, ed., "General Washington to Major Henry Lee, 22 October, 1780," *The Writings of George Washington*, Vol. 20, 240

> *Resolved, That the pressing emergency of our southern affairs requiring as speedy a reinforcement of cavalry as possible, Major Lee's corps be ordered to proceed immediately on their route to join the southern army.*[72]

A week after Lee was ordered southward, Congress, influenced by General Greene's intercession, granted the promotions that Major Lee had long sought for himself and his senior cavalry commander, Henry Peyton.[73] Hard luck Captain Allen McLane was denied his promotion (through no fault of Lee) and his resentment of Lee grew even stronger. This resentment probably contributed to McLane's departure from the legion in February 1781.

Newly promoted Lieutenant Colonel Lee wanted to accompany General Greene southward, but there was much to do to prepare his corps for the long march and General Greene could not wait. It would be two months before Lee and his legion managed to reunite with General Greene in South Carolina.

---

[72] Ford, ed., "31 October, 1780," *Journals of the Continental Congress*, Vol. 18, 997

[73] Ford, ed., "6 November, 1780," *Journals of the Continental Congress*, Vol. 18, 1023

Chapter Seven

"Heavy rains, deep creeks, bad roads, poor horses, and...want of provisions..."

On January 7, 1781, after a long march south that included stops in Virginia to recruit and better equip and outfit his men, Lieutenant Colonel Lee and his legion, 280 strong, joined the American southern army along the Pee Dee River in South Carolina.[1] Lee and his men were undoubtedly troubled at the condition in which they found the poorly clad and dispirited American southern army. Its commander, General Nathanael Greene, had himself arrived in the Carolinas a month earlier and had reported to Congress two weeks before Lee's arrival that his troops were *"so naked & destitute of every thing, that the greater part is rendered unfit for any kind of duty."*[2] Lee's men saw for themselves the poor condition of the army and probably wondered if they were destined for a similar fate.

The condition of his troops was not General Greene's only concern. He also lamented to Congress about his lack of cavalry, something that Lieutenant Colonel Lee's arrival helped remedy. Yet, despite an acute shortage of horsemen, Greene decided that Lee's Legion would best be utilized in support of Brigadier General Francis Marion along the Santee

---

[1] Lee, 223
[2] Richard K. Showman, ed., "General Greene to Congress, 28 December, 1780," *The Papers of General Nathanael Greene*, Vol. 7, 8

River and ordered Lee to unite with Marion and cooperate with his efforts there.

In the months following the fall of Charleston, General Marion had waged some of the only significant opposition to the British in South Carolina. Commanding parties of South Carolina militia that at times numbered less than 50 men and at other times were hundreds strong, Marion and his mounted militia raided British posts and their Tory allies along the Santee and Pee Dee Rivers. Coming in the wake of a series of American defeats in South Carolina, Marion's daring activities helped preserve a small degree of American resistance and morale and earned him the title of the Swamp Fox from future historians.

Lee arrived in Marion's camp on January 23rd and almost immediately discussed a plan to attack the 200 man British garrison at Georgetown on the upper coast of South Carolina. The plan called for the infantry of Lee's Legion under Captain Patrick Carnes to float down the Pee Dee River under cover of darkness to an island above the town. They were to conceal themselves and their rafts on the island all the following day. If everything went smoothly, Captain Carnes and his men would resume their water voyage to Georgetown the following night and arrive sometime after midnight.[3] They would then enter the town from the river, seize the garrison's commander, Lieutenant Colonel George Campbell, and wait for Lieutenant Colonel Lee and General Marion to charge into Georgetown with their mounted troops to overwhelm the rest of the British garrison.[4]

---

[3] Lee, 224
[4] Ibid.

South Carolina

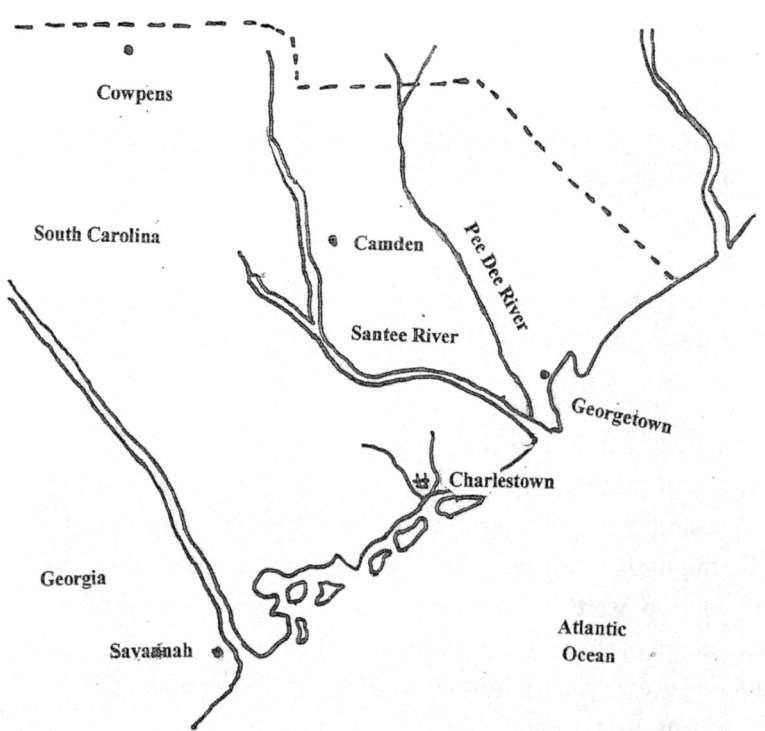

The attack initially went according to plan. Captain Carnes successfully eluded British patrols and pickets, landed in Georgetown on the second night and captured Lieutenant Colonel Campbell, while Lee and Marion stormed into town. The plan began to unravel, however, when Campbell's troops failed to come to his rescue. Lee explained what happened in his memoirs.

> *On the first fire which took place at the commandant's quarters, the militia of Marion and the dragoons of Lee rushed into the town, prepared to bear down all resistance. To the astonishment of these officers, every thing was quiet...Not a British soldier appeared; not one attempted to either gain the fort, or repair to the commandant. Having discovered their enemy, the troops of the garrison kept close to their respective quarters, barricaded the doors, and determined there to defend themselves.*[5]

Lee noted that his force lacked the necessary firepower or equipment to dislodge the enemy from the fort and houses and as a result, they were compelled to retire before sunrise after inflicting only a few casualties on the garrison.[6] Although Lee and Marion were disappointed in their raid on Georgetown, their efforts reinforced the growing view, highlighted by the stunning American victory at Cowpens a week earlier, that the Americans were not completely defeated or demoralized in South Carolina.

---

[5] Ibid.
[6] Ibid.

Cowpens

The decisive American victory at Cowpens occurred under Lieutenant Colonel Lee's old comrade from Valley Forge, Brigadier General Daniel Morgan. Morgan's heroics at Quebec in 1775 and Saratoga in 1777 had earned him recognition as a bold and courageous commander, but fatigue and discouragement at the lack of a promotion caused Morgan to resign from the army in 1779. The string of American defeats in South Carolina at Charlestown, the Waxhaws, and Camden in 1780 prompted Morgan to return to the army. Congress made him a brigadier general and ordered him to join the dispirited southern army in the Carolinas where he took command of the army's light corps.

In late December 1780, two weeks before Lieutenant Colonel Lee joined the southern army, General Greene sent Morgan's light corps westward to threaten the British outpost at Ninety-Six. General Morgan's force consisted of over 300 seasoned continentals, two companies of Virginia militia, and nearly 100 light dragoons under Lieutenant Colonel William Washington of Virginia, a distant cousin of General Washington.[7]

In response, British General Charles Cornwallis ordered Lieutenant Colonel Banastre Tarleton and his Legion to reinforce Ninety-Six. Tarleton and his men had developed a fierce reputation for their bold, (some would say brutal) tactics, and his legion of over 500 infantry and dragoons, augmented by three separate infantry detachments totaling 600 additional men, were an excellent counter-measure to Morgan's force.[8] Tarleton positioned his detachment to screen Ninety-Six, but soon realized that General Morgan did not

---

[7] Ibid., 222
[8] Lawrence E. Babits, *A Devil of a Whipping: The Battle of Cowpens*, (Chapel Hill : The University of North Carolina Press, 1998), 44-45

intend to strike the outpost. Never one to avoid a fight, Tarleton marched north towards Morgan. Aware of the threat General Cornwallis posed to the east, General Morgan withdrew northwest and halted at Cowpens to make a stand.

General Morgan had a knack for inspiring troops and spent the eve of battle preparing his men to fight. One soldier who witnessed Morgan's efforts recalled

> *It was upon this occasion I was more perfectly convinced of Gen. Morgan's qualifications to command militia, than I had ever before been. He went among the volunteers, helped them fix their swords, joked with them about their sweet-hearts, told them to keep in good spirits, and the day would be ours. And long after I laid down, he was going about among the soldiers encouraging them, and telling them that the old wagoner would crack his whip over Ben. [Tarleton] in the morning, as sure as they lived. 'Just hold up your heads, boys, three fires,' he would say, ' and you are free, and then when you return to your homes, how the old folks will bless you, and the girls kiss you, for your gallant conduct!' I don't believe he slept a wink that night!*[9]

General Morgan's battle plan involved a defense in depth comprising three battle lines. He placed approximately 150 militia riflemen in a skirmish line that extended across the road from which Tarleton was expected to approach.[10] The

---

[9] George Scheer and Hugh Rankin, "Account of Thomas Young," *Rebels and Redcoats: The American Revolution through the Eyes of Those Who Fought and Lived It*, (NY: Da Capo Press, 1987), 428
[10] Boatner III, 293

skirmishers deployed in loose order behind trees and were ordered to *"feel the enemy as he approached."*[11] When pressed, the riflemen were to withdraw to the flanks of the next American line 150 yards in their rear.[12]

Morgan's second line also stretched across the road but in a much tighter formation, the men standing nearly shoulder to shoulder in two ranks. This militia line consisted of 300 troops divided into four battalions.[13] General Morgan expected each battalion to fire at least two volleys before they retreated to the third and final American line.

Morgan's third line consisted of his best troops, Maryland, Delaware, and Virginia continentals under Lieutenant Colonel John Howard of Maryland. They were augmented by a few companies of seasoned militia from Virginia and North Carolina. This third line was formed about 150 yards behind the militia line and numbered about 550. In reserve, behind the third line, waited 120 continental and militia cavalry under Lieutenant Colonel Washington.[14] A few of Washington's horsemen were detached and posted three miles in advance of Cowpens to warn of Tarleton's approach.

Lieutenant Colonel Tarleton approached Cowpens early in the morning of January 17th. He commanded approximately 1,100 men, including over 300 cavalry and 750 infantry troops. Tarleton also had 2 three pound artillery pieces.[15]

---

[11] Babits, 81
[12] Ibid.
[13] Boatner III, 293
Note: The number of troops in the militia line has been reported as high as 1,000. See Tarleton, 216; If the bulk of the skirmish line joined the militia line then the number would have approached 500.
[14] Babits, 41-42
[15] John Buchanan, *The Road to Guilford Courthouse: The American Revolution in the Carolinas*, (NY : John Wiley & Sons, 1997), 321

Tarleton's force confronted Morgan's cavalry pickets before sunrise and pressed forward, reaching Cowpens at dawn. The Americans were deployed and waiting for him. Thomas Young, a militiaman attached to Washington's cavalry, recalled,

> *The morning...was bitterly cold...About sunrise, the British line advanced at a sort of trot with a loud halloo. It was the most beautiful line I ever saw. When they shouted, I heard Morgan say, 'They give us the British halloo, boys. Give them the Indian halloo, by G--!' and he galloped along the lines, cheering the men and telling them not to fire until we could see the whites of their eyes.*[16]

Lieutenant Colonel Tarleton's dragoons screened his main body and engaged Morgan's riflemen in the skirmish line but failed to dislodge them. Undeterred, Tarleton ordered his entire force forward. His troops dropped their extra gear and formed a battle line across the road. As they advanced in open order towards the American riflemen they were peppered with "*a heavy & galling fire.*"[17] A few British soldiers responded with unauthorized shots, but most held their fire and continued forward, driving the riflemen back to the militia line.

As Tarleton's force approached within 50 yards of the next American line, each militia battalion unleashed a well aimed volley into their ranks. A British officer noted that, "*The effect of the fire was considerable, it produced something like a recoil.*"[18] Despite the heavy fire, Tarleton's men pressed on

---

[16] Scheer and Rankin, " Thomas Young", 430
[17] Showman, ed., "General Morgan to General Greene, 19 January, 1781," *The Papers of General Nathanael Greene*, Vol. 7, 154
[18] Ibid., 92

and forced the militia to retreat before most had a chance to fire a second time. As the militia withdrew to the third line they were suddenly attacked by Tarleton's cavalry. Only the timely arrival of Lieutenant Colonel Washington's horsemen saved the panic stricken militia from disaster.

While Washington engaged Tarleton's cavalry on the American left flank, Morgan's men on the third line braced to engage Tarleton's infantry. The British troops briefly halted to dress and close ranks and then resumed their attack. General Morgan proudly noted that the approaching enemy was, *"received [by] a well-directed and incessant Fire,"* from the continentals.[19] The British responded in kind. Tarleton observed, *"The fire on both sides was well supported, and produced much slaughter."* [20] Another participant noted that both sides *"Maintained their ground with great bravery; and the conflict...was obstinate and bloody."*[21]

Lieutenant Colonel Tarleton tried to break the deadlock by striking at Morgan's right flank. This key position in the American line was held by a company of Virginia continentals under Captain Andrew Wallace. They were supported by a party of North Carolina riflemen along the woods to their right. Tarleton sent in his cavalry and reserve battalion of infantry and hit the American right flank hard.

---

[19] Showman, ed., "General Morgan to General Greene, 19 January, 1781," *The Papers of General Nathanael Greene,* Vol. 7, 154

[20] Lieut. Col. Banastre Tarleton, " Tarleton to Earl Cornwallis, 4 January, 1781," *A History of the Campaigns of 1780 and 1781 in the Southern Provinces of North America,* (North Stratford, NH : Ayer Company Publishers, 1999), 216
(Originally published in 1787)

[21] Babits, 103

Tarleton's dragoons scattered the American riflemen and were about to gain Captain Wallace's rear when Lieutenant Colonel William Washington's cavalry rushed onto the scene and stopped them cold. Although Tarleton's horsemen were forced to retire, his infantry still threatened Morgan's right flank so Lieutenant Colonel John Howard, the commander of the continental troops, ordered an adjustment in the American line. Unfortunately, Howard's orders were misunderstood and some confusion occurred in their execution. He explained after the battle:

> *Seeing my right flank was exposed to the enemy, I attempted to change the front of Wallace's company. In doing this, some confusion ensued, and first a part and then the whole of the company commenced a retreat. The officers along the line seeing this and supposing that orders had been given for a retreat, faced their men about and moved off.*[22]

General Morgan was initially upset by the unauthorized withdrawal, but Lieutenant Colonel Howard assured him that the battle was not lost.[23] Morgan recalled that,

> *We retired in good Order about 50 Paces, formed, advanced on the Enemy & gave them a fortunate Volley which threw **them** into Disorder.*[24]

---

[22] Commager and Steele, "Lieutenant Colonel John Eager Howard's Account," *The Spirit of Seventy-Six*, 1156

[23] Ibid.

[24] Showman, ed., "General Morgan to General Greene, 19 January, 1781," *The Papers of General Nathanael Greene*, Vol. 7, 154

In truth, Tarleton's men were disordered prior to the American volley. The initial withdrawal of the American third line prompted the British to rush forward. *"They are coming on like a mob,"* Lieutenant Colonel Washington told Howard, *"give them a fire and I will charge them."*[25] Howard did precisely that, ordering his men to face about. *"In a minute we had a perfect line,"* recalled Howard.[26] He continued,

> *The enemy were now very near us. Our men commenced a very destructive fire, which they little expected, and a few rounds occasioned great disorder in their ranks. While in this confusion, I ordered a charge with the bayonet, which order was obeyed with great alacrity.*[27]

General Morgan proudly described the impact of the American bayonet charge to General Greene:

> *Lt. Colonel Howard observing* [the enemy's disorder] *gave orders for the Line to charge Bayonets, which was done with such Address that they fled with the utmost Precipitation, leaving the Field Pieces in our Possession. We pushed our Advantage so effectually, that they never had an Opportunity of rallying.*[28]

Tarleton's defeat at Cowpens was nearly total. His detachment was shattered, and he lost over 800 men.[29] Only Tarleton and his horsemen escaped.

---

[25] Babits, 117
[26] Commager and Steele, "Lieutenant Colonel John Eager Howard's Account," *The Spirit of Seventy-Six*, 1156
[27] Ibid. 1157
[28] Showman, ed., "General Morgan to General Greene, 19 January, 1781," *The Papers of General Nathanael Greene*, Vol. 7, 154
[29] Babits, 143

Morgan's Retreat

General Morgan realized that his light corps was still in danger from General Cornwallis and the main British army to the east, so he immediately commenced a march northward. Morgan's objective was two-fold, to move the hundreds of British prisoners out of reach of Cornwallis, and to reunite with General Greene as fast as possible.

As Morgan marched into North Carolina, most of the Georgia and South Carolina militia that were with him departed. On January 23rd, the same day that Lieutenant Colonel Lee joined General Marion and prepared to attack Georgetown, Morgan's force crossed the Catawba River. The British prisoners continued on towards Virginia under the guard of the Virginia militia, who were themselves returning home. That left General Morgan with only 300 continentals and a handful of North Carolina militia, yet Morgan made the bold decision to stand his ground at the Catawba, await instructions from General Greene, and hope for militia reinforcements.[30] Morgan guarded the fords of the Catawba River for over a week waiting for Cornwallis to advance.

Hampered by a large baggage train and heavy rain that flooded the creeks and rivers and turned the roads into a quagmire, Cornwallis's pursuit proceeded at a crawl. In frustration, he took the extreme measure of destroying most of his baggage and wagons and impatiently waited for the flooded Catawba River to recede. When it finally did so on January 31st, Cornwallis advanced.

---

[30] Showman, ed., "General Daniel Morgan to General Nathanael Greene, 23 January, 1781," *The Papers of General Nathanael Greene,* Vol. 7, 178

Defending the crossing points against Cornwallis's 2,500 man army were Morgan's 300 continentals and 800 North Carolina militia who had arrived during the week under General William Davidson.[31] A large detachment of these men, led by General Davidson himself, fiercely resisted a late night British crossing at Cowans Ford. General Cornwallis described the crossing in a letter to Lord Germain.

> *Full of confidence in the zeal and gallantry of Brigadier-general O' Hara* [and his men] *I ordered them to...* [cross the river and] *not to fire until they gained the opposite bank. Their behavior justified my high opinion of them; for a constant fire from the enemy, in a ford upwards of five hundred yards wide, in many places up to their middle, with a rocky bottom and strong current, made no impression on their cool and determined valour, nor checked their passage.*[32]

Davidson's men inflicted scores of casualties on the British, but failed to halt their crossing. As the British reached the other side of the river, General Davidson's militia fled eastward. They did so without their commander, who was killed at the end of the engagement.

The British crossing at Cowans Ford left General Morgan, who had been joined a day earlier by General Greene and a tiny escort of dragoons, with no option but to withdraw northeast through Salisbury and across the Yadkin River.

---

[31] Showman, ed., "General Daniel Morgan to General Nathanael Greene, 29 January, 1781," *The Papers of General Nathanael Greene*, Vol. 7, 215

[32] Tarleton, "Cornwallis to Lord Germain, 17 March, 1781," 262

North Carolina

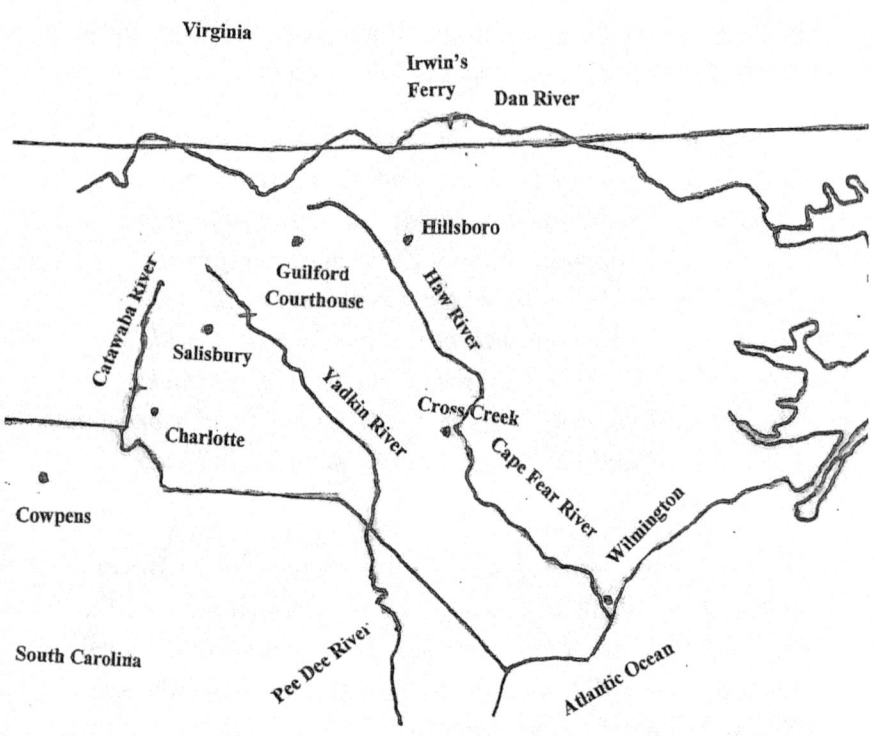

Three days earlier, General Greene had abruptly left the main army on the Pee Dee River (one hundred miles to the east) and rushed westward to confer with General Morgan. Upon his arrival, Greene sent word back to General Isaac Huger to send the army's heavy baggage northward to Guilford Court House and march the rest of the army towards Salisbury to unite with Morgan. He also instructed Huger to order Lieutenant Colonel Lee and his legion (who was still attached to General Marion) to join Morgan's corps as quickly as possible.

As Lieutenant Colonel Lee and General Huger pushed their men northwestward, a bit of good fortune shined on Morgan and his outnumbered men. Pursued by Cornwallis's army, Morgan's troops crossed the Yadkin River just before heavy rains flooded it and stranded Cornwallis on the other side. General Morgan halted his retreat and impatiently waited on the east bank of the river for reinforcements. While he did so, Morgan grudgingly addressed a problem that had plagued him for weeks. Exhausted and in extreme pain due to sciatica, General Morgan relinquished command of the light corps and returned to Virginia to recuperate. General Greene took charge of Morgan's men and decided they were too few to defend the receding Yadkin River so he marched north towards Guilford Court House to unite with the main body of the American army, which was also directed to march to Guilford Court House. The two forces, including Lee's Legion, finally reunited on February 9th.

General Greene immediately held a war council with his ranking officers where it was revealed that the entire army comprised only 1426 ill-clad and poorly armed continentals and 600 poorly armed militia.[33] Approximately 25 miles west of Guilford Court House was General Cornwallis and 2,500 enemy troops eager to catch the Americans and avenge Cowpens. General Greene and his officers agreed that it would be best to avoid an engagement with Cornwallis until they were reinforced by more militia, so Greene ordered a retreat to the Dan River in Virginia. General Greene instructed the light infantry corps, now under the command of Colonel Otho Williams of Maryland, to screen the army during its retreat northward. Colonel Williams was a veteran officer and an excellent choice to command the light corps in the absence of General Morgan. Before he departed for the Dan River with the bulk of the army, General Greene wrote to General Washington and updated him on the situation.

> *Since I wrote your Excellency* [last] ... *Lord Cornwallis has been constantly in pursuit of the Light Infantry and the prisoners...and* [is] *still pushing into the country with great rapidity. The moment I was informed of the movements of Lord Cornwallis I put the army in motion on Pedee and...set out to join the Light Infantry in order to collect the Militia and embarrass the enemy 'till we could effect a junction of our forces.... Heavy rains, deep creeks, bad roads, poor horses and broken harnesses as well as delays for want of provisions prevented our forming a junction as early as I expected, and fearing that the river might fall so as to be fordable, I ordered the*

---

[33] Showman, ed., "War Council 9 February, 1781," *The Papers of General Nathanael Greene*, Vol. 7, 261

army to file off to this place.... [Guilford Court House].[34]

General Greene expressed disappointment in the militia turnout and presented a bleak picture of the army.

> *We have no provisions but what we receive from our daily collections. Under these circumstances I called a council who unanimously advised to avoid an action and to retire beyond the* [Dan River] *immediately.... I have formed a light army composed of the cavalry...and the Legion* [Lee's] *amounting to 240, a detachment of 280 Infantry under Lt. Col. Howard, the Infantry of Lt. Col. Lee's legion and 60 Virginia Rifle Men making in their whole 700 Men which will be ordered with the Militia to harass the enemy in their advance, check their progress and if possible give us an opportunity to retire without a general action.*[35]

---

[34] Showman, ed., "General Greene to General Washington, 9 February, 1781," *The Papers of General Nathanael Greene,* Vol. 7, 267-269
[35] Ibid.

Race to the Dan

While General Greene and the main body of the American army prepared to retreat northeast towards the Dan River, Lieutenant Colonel Lee and his legion marched northwest of Guilford Courthouse with Colonel Otho Williams's 700 man light corps. General Greene had made arrangements to gather boats in the vicinity of Irwin's and Boyd's ferries on the Dan River in Virginia, but he needed time to reach the river and transport his army and its baggage across. With Cornwallis no longer encumbered by a large baggage train and only 25 miles west of Guilford Court House, Greene's army was in real jeopardy of being overtaken.

Colonel Williams and the American light corps had one mission, to protect the main body of the army long enough to allow it to escape across the Dan River. Of course, it would be most desirable if Williams could achieve his mission without sacrificing the light corps, but the American light troops realized that they were expendable if it meant saving the rest of the American army. As things turned out, this was almost the case.

While General Greene and the main body of the army marched northeast, Colonel Williams and the light corps was sent northwest to find Cornwallis and delay his pursuit. General Cornwallis had mistakenly assumed that the Americans would attempt to cross the Dan River via the upper fords northwest of Guilford Courthouse and positioned his army accordingly.[36] When he realized that he was wrong, Cornwallis directed his march towards Dix's Ferry (modern

---

[36] Tarleton, "Earl Cornwallis to Lord George Germain, 17 March, 1781," 264

day Danville, Virginia). This crossing was twenty miles to the west of General Greene's true destination of Irwin's and Boyd's Ferries and precisely the route that Colonel Williams hoped to draw Cornwallis down.

The two sides made contact with each other on the morning of February 11th. Warned by a local Whig that the enemy was nearby, Colonel Williams ordered Lieutenant Colonel Lee to investigate.[37] Lee's men were having breakfast when the order arrived so he sent only one troop of cavalry under Captain James Armstrong southwestward and ordered the rest of his legion to follow after they had finished breakfast.[38] Lee accompanied Armstrong, but after a few miles decided that the report was mistaken. He halted and ordered Captain Armstrong to remain with three dragoons and the local Whig while Lee led the rest of the troop back to rejoin the legion.[39]

The civilian guide with Lee expressed great anxiety at remaining behind on his tired horse so Lee ordered his young bugler, James Gillies, to exchange horses with the man. He then sent Gillies back to the light infantry camp to update Colonel Williams.[40]

As Lee and his troopers slowly retraced their route back, they heard gunfire in their rear where they had left Captain Armstrong. Lee posted the dragoons with him off the road in the woods and waited. Within moments Captain Armstrong and his party came charging past, closely pursued by a troop of enemy cavalry under Captain Patrick Miller of Tarleton's

---

[37] Showman, ed., "Colonel Otho Williams to General Greene, 11 February, 1781," *The Papers of General Nathanael Greene*, Vol. 7, 283

[38] Babits and Howard, 32

[39] Lee, 240

[40] Ibid.

Legion. Neither noticed Lieutenant Colonel Lee and his horsemen.

Lee led his party of dragoons in pursuit of Miller, who was unable to catch Captain Armstrong but overtook James Gillies on the worn out horse. In a matter of moments the young, unarmed bugler was unhorsed and sabered mercilessly. Lee and his party arrived in time to witness the attack and charged headlong into Miller and his men. Infuriated by what they considered cold blooded murder, they decimated Miller's troop, killing 18 and capturing the captain and a few others.[41] Lieutenant Colonel Lee angrily reprimanded his troopers when they brought back prisoners; he had instructed that no quarter be given in retaliation for what happened to James Gillies.[42] Lee recalled in his memoirs years later that,

> *Miller, being peremptorily charged with the atrocity perpetrated in his view, was told to prepare for death. The captain, with some show of reason, asserted that intelligence being his object, it was his wish and interest to save the soldier; that he had tried to do so; but his dragoons being intoxicated, all his efforts were ineffectual. He added, that in the terrible slaughter* [of the Americans at the Waxaws] *his humanity was experienced, and had been acknowledged by some of the Americans who escaped death on that bloody day. Lee was somewhat mollified by this rational apology, and was disposed to substitute one of the prisoners; but soon overtaking the speechless, dying youth...when able to*

---

[41] Ibid., 241
[42] Ibid., 242

> speak, confirmed his former impressions, he returned with unrelenting sternness to his first decision. Descending a long hill, he repeated his determination to sacrifice Miller in the vale through which they were about to pass; and handing him a pencil desired him to note on paper whatever he might wish to make known to his friends, with an assurance that it should be transmitted to the British general.[43]

Captain Miller was saved by the approach of the main British column which forced Lee to abandon Miller's execution and rush the prisoners to the rear.

With Cornwallis now on the trail of the American light corps, Colonel Williams hastened his march towards Dix's Ferry. Every mile of road that Colonel Williams drew Cornwallis down provided General Greene with more time to reach the Dan River. The danger for the American light troops, however, was that every step also took them further away from the support of General Greene's army. By midday Colonel Williams decided that he had provided General Greene with an adequate head start to the Dan, so he redirected his march towards Irwin's Ferry to join him there.[44]

Lee's Legion, as usual, assumed the responsibilities of the rear guard. At some point in the march Lee diverted his men onto an obscure path along the route taken by the light corps. He hoped to provide his men with a little rest and then use the path to catch up to the light corps. After a short time on this route Lee encountered a generous farmer who willingly

---

[43] Ibid.
[44] Ibid.

provided forage for Lee's horses and provisions for his men. Lee recorded in his memoirs what happened next.

> *The obscurity of the narrow road taken by Lee lulled every suspicion with respect to the enemy; and a few vedettes only were placed at intermediate points, rather to give notice when the British should pass along, than to guard the Legion from surprise. This precaution was most fortunate; for it so happened that Lord Cornwallis, having ascertained that Greene had directed his course to Irwin's Ferry, determined to avail himself of the nearest route to gain the road of his enemy, and took the path which Lee had selected. Our horses were unbridled, with abundance of provender before them; the hospitable farmer had liberally bestowed his meal and bacon.... To the surprise and grief of all, the pleasant prospect was instantly marred by the fire of the advanced vedettes – certain signal of the enemy's approach. Before the farm was a creek, which, in consequence of the late incessant rains, could be passed only by a bridge, not more distant from the enemy then from our party. The cavalry being speedily arrayed, moved to support the vedettes; while the infantry were ordered, in full run, to seize and hold the bridge.*
>
> *The enemy was equally surprised with ourselves at this unexpected meeting; and the light party in front halted, to report and be directed. This pause was sufficient. The bridge was gained, and soon passed by the corps of Lee. The British followed.*

> *The road over the bridge leading through cultivated fields for a mile, the British army was in full view of the troops of Lee as the latter ascended the eminence, on whose summit they entered the great road to Irwin's Ferry.*[45]

Now the race was really on!

With General Cornwallis pressing Lee and the American light corps, it was crucial that Lieutenant Colonel Lee and his men remain vigilant. Cornwallis's advance troops occasionally came within musket shot of Lee's rear guard and the American light corps literally drove themselves to exhaustion to stay ahead of the enemy. The Americans halted only when the British did. Lieutenant Colonel Lee recalled in his memoirs that,

> *The duty, severe in the day, became more so at night; for numerous patrols and strong pickets were necessarily furnished by the light troops, not only for their own safety, but to prevent the enemy from placing himself by a circuitous march, between Williams and Greene. Such a maneuver would have been fatal to the American army; and, to render it impossible, half of the troops were alternately appointed every night to duty: so that each man, during the retreat, was entitled to but six hours repose in forty-eight.*[46]

Such hard duty took a toll on the men and horses, yet, the American light troops marched on. At one point it appeared

---

[45] Ibid., 244
[46] Ibid., 238

that their efforts to screen the main body had failed when the van of the light corps spotted numerous campfires off in the distance. With Cornwallis still pressing their rear, Lee and the light corps worried that they had caught up to General Greene. If this were true, then all of their effort and sacrifice to provide General Greene with time to escape had been wasted. As they approached the campfires the exhausted light troops resigned themselves to one last doomed engagement against a vastly superior enemy.[47] Lieutenant Colonel Lee recalled that

> *No pen can describe the heart-rendering feelings of our brave and wearied troops.... Our dauntless corps were convinced that the crisis had now arrived when its self sacrifice could alone give a chance of escape of the main body. With one voice was announced the noble resolution to turn on the foe, and by dint of desperate courage, so to cripple him as to force a discontinuance of pursuit.[48]*

However, to the great relief of the light troops, it was soon discovered that Greene's troops had left the campsite hours earlier and that local militia had maintained the fires for the benefit of the light corps.[49] Unfortunately for Colonel Williams and his tired men, the close proximity of General Cornwallis prevented the American light troops from enjoying the warmth of the campfires. Colonel Williams did find the time to write to General Greene, however, and express his anxiety about the slow progress of Greene's main body of troops.

---

[47] Ibid., 245
[48] Ibid.
[49] Ibid.

> *I was exceedingly concern'd to hear...that you were yet 25 miles from the ferry. My Dear General at Sun Down the Enemy were only 22 miles from you and may be in motion now or will most probably [be] by 3 oClock in the morning.... Rely on it, my Dear Sir, it is possible for you to be overtaken before you can cross the Dan even if you had 20 boats.... I conclude you march'd as far today as you could and if your Army can make but Eleven miles in a Day you will not be able to pass the ferry in less than two Days more. In less time than that we will be driven in to your Camp or I must risqué the Troops I've the Honor to command and in doing so I risqué every thing.... The Gentlemen of Cavalry assure me their Horses want refreshment exceedingly and our Infantry are so excessively fatigu'd that I'm confident I lose men every Day. We have been all this Day almost in presence of the Enemy but have sustain'd no loss but of Sick and Strollers.*[50]

Despite Colonel Williams's pessimism, the Americans marched on. Lieutenant Colonel Lee recalled that

> *About midnight our troops were put in motion, in consequence of the enemy's advance on our pickets, which the British general had been induced to order, from knowing that he was within forty miles of the Dan, and that all his hopes depended on the exertions of the following day. Animated with the prospect of soon terminating their present labors, the light troops*

---

[50] Showman, ed., "Col. Otho Williams to Gen. Greene, 13 February 1781," *The Papers of General Nathanael Greene*, Vol. 7, 285-286

> resumed their march with alacrity. The roads continued deep and broken, and were rendered worse by being incrusted with frost: nevertheless, the march was pushed with great expedition.[51]

General Cornwallis described his pursuit in similar terms.

> *Nothing could exceed the patience and alacrity of the officers and soldiers under every species of hardship and fatigue, in endeavouring to overtake* [the Americans]: *But our intelligence upon this occasion was exceedingly defective; which, with heavy rains, bad roads, and the passage of many deep creeks, and bridges destroyed by the enemy's light troops, rendered all our exertions vain.*[52]

Good news finally reached the American light troops on February 14th in the form of a series of messages from General Greene informing Colonel Williams of the army's arrival at the Dan River. Greene wrote at 2:00 p.m. that, *"The greater part of our wagons are over and the troops are crossing,"* and at 5:30 p.m. Greene announced, *"The stage is clear."*[53]

The good news inspired the light troops to march with renewed vigour and they reached the ferry crossings hours ahead of the British. General Greene greeted Colonel Williams at the riverbank and crossed to the north side with him. Lieutenant Colonel Lee arrived with his Legion soon after and proudly recalled in his memoirs that he was the last

---

[51] Lee, 246

[52] Tarleton, "Cornwallis to Lord Germain 17 March, 1781," 264

[53] Showman, ed., "Col. Otho Williams to Gen. Greene, 14 February 1781," *The Papers of General Nathanael Greene*, Vol. 7, 287

to cross the Dan River.[54] The American army had achieved its goal, it had crossed the Dan River intact. It remained to be seen whether they had escaped General Cornwallis's wrath or only postponed it.

---

[54] Lee, 247

Chapter Eight

"Unless our Army is greatly reinforced I see nothing to prevent [the enemy's] future progress"

The days immediately following the American army's escape across the Dan River were anxious ones for General Greene and his troops. He was not convinced that his army was out of danger and informed Governor Thomas Jefferson of Virginia that he expected Cornwallis to cross the river as soon as he was able and force a battle.

> *Our Army is so great an object with Lord Cornwallis, and the reduction of the Southern States depend so much upon it, that I think it highly probable he will push us still farther; and I am not without my apprehensions that he will oblige us to risqué a general action, however contrary to our wishes or interest, which cannot fail to prove ruinous to the Army as our force is but little more than one half the enemy's, and there are few or no Militia that have joined us on our March, or at least not more than one or two hundred.... Unless our Army is greatly reinforced I see nothing to prevent [the enemy's] future progress.*[1]

---

[1] Julian Boyd, ed., "General Greene to Governor Jefferson, 15 February, 1781," *The Papers of Thomas Jefferson*, Vol. 4, 615

General Greene gave an equally bleak assessment of his situation to General Washington.

> *Lord Cornwallis has been at our heels from day to day ever since we left Guilford and our movements from thence to this place have been of the most critical kind, having a river in our front, and the enemy in our rear. But happily we have crossed without the loss of either men or Stores.... The enemy are on the other side of the river and as it is falling, I expect it will be fordable before night; and the fords are so numerous, and the enemy lay in such an advantageous situation for crossing that it would be a folly to think of defending them, as it would reduce our force to small parties which might prove our ruin. The miserable situation of the troops for want of clothing has rendered the march the most painful imaginable, several hundreds of the Soldiers tracking the ground with their bloody feet.... The enemy's movements have been so rapid and the Country under such terror that few or no Militia have joined us, and the greater part we had have fallen off.... Nothing is yet done to give me effectual support and I am not a little apprehensive that it is out of the power of Virginia & North Carolina to afford it.*[2]

General Greene's apprehension significantly eased when intelligence arrived the next day of General Cornwallis's intentions to march to Hillsborough, North Carolina. Greene was further encouraged by the arrival of long awaited militia

---

[2] Showman, ed., "General Greene to General Washington, 15 February, 1781," *The Papers of General Nathanael Greene*, Vol. 7, 293-94

reinforcements and reports of even larger numbers of militia en route to join the army. Just two days after his pessimistic letters to Jefferson and Washington, General Greene wrote that he was trying to tempt General Cornwallis into actually crossing the Dan.

> *Our Army is on the North side of the Banister River, encamped at Halifax Court House in Virginia, in order to tempt the Enemy to cross the* [Dan] *River, as the most pleasing prospect presents itself of a strong reinforcement from the Militia of this State.*[3]

On February 19th, Greene ordered General Andrew Pickens, who was still in North Carolina with approximately 700 militia troops, to harass and delay General Cornwallis on his march to Hillsborough.[4] Greene hoped to turn the tables on Cornwallis, re-cross the Dan River, and attack Cornwallis with his reinforced army. Lieutenant Colonel Lee and his Legion re-crossed the river ahead of Greene's army to pursue their former pursuers. Lee informed Greene on February 20th, that he was only twenty-five miles from Hillsborough and that advanced parties of his horsemen had reached the enemy's lines.[5]

Although General Greene and the bulk of his army had yet to re-cross the Dan by February 21st, (two days after Lee crossed) Greene's new found confidence in his army was still evident when he wrote to General Pickens.

---

[3] Showman, ed., "General Greene to Joseph Clay, 17 February, 1781," *The Papers of General Nathanael Greene,* Vol. 7, 300
[4] Showman, ed., "General Greene to General Pickens, 19 February, 1781," *The Papers of General Nathanael Greene,* Vol. 7, 316
[5] Showman, ed., "Lt. Col. Lee to Gen. Greene, 20 February, 1781," *The Papers of General Nathanael Greene,* Vol. 7, 324

> *Lt Col Lee with his Legion is in full pursuit. Col Williams with the light Infantry is also on the march. The army will cross the river in the Morning, with a considerable reinforcement of Militia. If we can get up with the enemy, I have no doubt of giving a good account of them.*[6]

General Greene's enthusiasm to attack Cornwallis cooled considerably the following day, however, when he learned that a large reinforcement of riflemen would not be joining him as expected. General Greene halted the main army and ordered his three advance detachments under General Pickens, Lieutenant Colonel Lee and Colonel Williams, to watch for opportunities to harass the British and their Tory supporters.

Pyle's Hacking

Lee and Pickens joined forces on February 23rd and learned two days later that a British detachment under Banastre Tarleton was nearby. Tarleton's force of 200 cavalry, 150 British infantry, and 100 German jagers had been sent out from Hillsborough to protect a detachment of Tory militia that had formed under Colonel John Pyle.[7] Lee and Pickens immediately set out to attack Tarleton.

Their first attempt to catch Tarleton failed when the Americans rushed into Tarleton's vacated campsite hours after he had left, but two enemy stragglers were apprehended and the information they provided encouraged Lee and Pickens to continue. The American commanders decided to use a bit of

---

[6] Showman, ed., "General Greene to General Pickens, 21 February, 1781," *The Papers of General Nathanael Greene,* Vol. 7, 327
[7] Tarleton, 231

deception on their next attempt and pass themselves off as reinforcements from Hillsborough for Tarleton. This bold plan was possible because Lee's men were dressed in short green coats and leather caps that looked remarkably similar to Tarleton's Legion. Since General Pickens men were not dressed that way they had to trail behind Lee and stay out of sight in order for the ruse to be successful.

The scheme was soon tested when the advance guard of Lee's Legion encountered mounted scouts from Colonel Pyle's detachment of Tories. Lee recalled in his memoirs that Pyle's scouts were completely deceived and even believed that Lee was Tarleton.[8] The scouts returned to Colonel Pyle with a request from Lee to clear the road for "Tarleton's" detachment. Lee claimed in his memoirs that since Tarleton was his primary objective he planned to avoid a fight with Pyle's Tories. Instead, he planned to

> *Make known to* [Colonel Pyle] *his real character as soon as he should confront him, with a solemn assurance of his and his associates perfect exemption from injury, and with the choice of returning to their homes, or of taking a more generous part, by uniting with the defenders of their common country against a common foe.*[9]

As it turned out, Lee was unable to offer either of these options to Colonel Pyle. Instead, a bloody, one sided fight erupted. Lee, writing in the third person, recalled the incident in his memoirs.

---

[8] Lee, 256
[9] Ibid. 257

> *Lee passed along the line at the head of the column with a smiling countenance, dropping, occasionally, expressions complimentary to the good looks and commendable conduct of his loyal friends. At length he reached Colonel Pyle, when the customary civilities were promptly interchanged. Grasping Pyle by the hand, Lee was in the act of consummating his plan, when the enemy's left, discovering Pickens's militia, not sufficiently concealed, began to fire upon the rear of the cavalry commanded by Captain Eggleston. This officer instantly turned upon the foe, as did immediately the whole column. This conflict was quickly decided, and bloody on one side only.*[10]

Captain Joseph Graham, the commander of a company of mounted North Carolina militia, attributed the engagement to a different cause. He recalled that he and his men initially believed Pyle's men to be comrades, but soon realized their mistake.

> *I riding in front of the Militia dragoons, near to Capt. Eggleston who brought up Lee's rear, at a distance of forty or fifty yards, pointed out to him, the strip of red cloth on the hats of Pyle's men, as the mark of Tories. Eggleston appeared to doubt this, until he came nearly opposite to the end of their line, when riding up to the man on their left, who appeared as an officer, he inquired, "Who do you belong to?" The answer was promptly given, "To King George," upon which Eggleston struck him on the head with*

---

[10] Ibid., 258

> *his sword. Our [militia] dragoons well knew the red cloth on the hats to be the badge of Tories, but being under the immediate command of Lee, they had waited for orders. But seeing the example set by this officer, without waiting for further commands, they rushed upon them like a torrent. Lee's men, next to the [front] discovering this, reined in their horses to the right upon the Tory line, and in less than one minute the engagement was general.*[11]

At the loss of not a single man, Lee and Pickens destroyed Colonel Pyle's detachment. Estimates of Tory casualties vary, but it is clear that approximately 100 Tories were killed and an equal number were wounded, many grievously by the Legion's sabers. One important factor in the lopsided outcome was the continued belief among many of Pyle's men (even as they were being cut down) that Lee's Legion was actually Tarleton's Legion. Instead of fighting back, many Tories exclaimed their support for the King, thinking that this would halt the attack. Instead, they were struck down.[12]

Although the engagement was a decisive and victorious affair for the Americans, it was a fight that Lieutenant Colonel Lee had wished to avoid. Lee confirmed this in a letter to General Greene on the evening of the engagement.

---

[11] Major William Graham, "Joseph Graham to Judge Murphey, 20 December, 1827," *General Joseph Graham and His Papers on North Carolina Revolutionary History*, (Raleigh, NC: Edwards & Broughton, 1904), 207

[12] Graham, 319-320

> *The Legion cavalry passed* [Pyle's men] *agreeable to order, as if British troops. I did this, that no time might be lost in reaching Col Tarleton. The enemy discovered their mistake on the near approach of our militia & commenced action.*[13]

General Pickens, who was near the rear of the American force, also viewed the incident negatively and considered it a missed opportunity to surprise Tarleton.

> *Never was there a more glorious opportunity of cutting off a detachment* [Tarleton's] *than this when...our sanguine expectations were blasted by our falling in with a body of from two to three hundred Tories, under the command of a Colonel Piles...they suffered Colonel Lee's Horse to pass equal with their front. Our men were in some measure under the same mistake* [unaware that Pyle's men were the enemy] *but soon found out, and nigh one hundred were killed and the greatest part of the others wounded.*[14]

Although this one-sided affair cost Lieutenant Colonel Lee and General Pickens a chance to surprise Banastre Tarleton, the brutality of the fight and its outcome had a chilling effect on future Tory support for the British in North Carolina. Furthermore, Lee and Pickens had not given up on attacking Tarleton. Reinforced at the end of the day by hundreds of

---

[13] Showman, ed. "Lt. Col. Lee to General Greene, 25 February, 1781," *The Papers of General Nathanael Greene*, Vol. 7, 348
[14] Showman, ed., "General Pickens to General Greene, 26 February, 1781," *The Papers of General Nathanael Greene*, Vol. 7, 355

newly arrived Virginia riflemen under Colonel William Preston, Lee's men spent the night within a few miles of Tarleton. They hoped to engage Tarleton at daybreak, but he stole an early morning march on the Americans and returned to Hillsborough unscathed.

General Cornwallis and his re-united army of over 2,000 men departed Hillsborough the same day and marched west to obtain more plentiful provisions and protect the King's friends (Tories) between the Haw and Deep Rivers.[15]

General Greene, who was north of the Haw River with the main body of the American army, responded cautiously to Cornwallis's movements. He led his troops southward and linked up with Colonel Williams and the light corps along the Haw River.[16] Lee and Pickens joined General Greene on the Haw River and were re-attached to the light corps. Greene pushed the army across the Haw River the next day and sent the light corps further south across Reedy Fork Creek to screen the army and locate the enemy. Colonel Williams, in turn, sent Lieutenant Colonel Lee and his legion, as well as a detachment of Colonel Preston's riflemen and parties of North Carolina militia even further south to patrol along Alamance Creek.[17] On March 2nd, Lee's detachment engaged a large enemy party in a brief but heated skirmish. A much larger engagement involving Lee and the entire American light corps erupted four days later at Weitzel's Mill along Reedy Fork Creek.

---

[15] Tarleton, 234

[16] Showman, ed., "General Greene to General Steuben, 29 February, 1781," *The Papers of General Nathanael Greene,* Vol. 7, 375

[17] Lawrence E. Babits and Joshua B. Howard, *Long, Obstinate, and Bloody: The Battle of Guilford Courthouse,* (Chapel Hill, NC: The University of North Carolina Press, 2009), 43

Weitzel's Mill

The engagement at Weitzel's Mill was prompted by General Cornwallis's attempt to catch the American light corps by surprise. On the evening of March 5th, Cornwallis was informed that the American light corps, *"was posted carelessly at separate plantations for the convenience of* [foraging]."[18] Cornwallis marched his army towards the American light troops early the next morning hoping to, *"drive them in and...attack General Greene's* [main body] *if an opportunity offered."*[19] Greene had moved his troops back across the Haw River and was relatively safe, but Colonel Williams and his light troops were much closer to Cornwallis and unaware of the threat. Fortunately for the Americans, Cornwallis's movement was detected by an American patrol in time to warn the light corps. Colonel Williams recalled,

> [The enemy] *had approach'd within two miles of our position, and their intention was manifestly to surprise us. I immediately order'd the Troops to march to* [Weitzel's] *Mill and soon after was inform'd by two Prisoners that the Enemy were marching for the same place on a road parallel to that in which we were.*[20]

---

[18] Tarleton, "Earl Cornwallis to Lord George Germain, March 17, 1781," 266
[19] Ibid.
[20] Showman, ed., "Col. Otho Williams to General Greene, 7 March, 1781," *The Papers of General Nathanael Greene*, Vol. 7, 407

As the Americans and British raced to Weitzel's Mill to cross Reedy Fork Creek, Colonel Williams ordered Lee's Legion, Lieutenant Colonel William Washington's cavalry, and the riflemen under Colonel Campbell and Colonel Preston to slow the enemy's march. Williams reported that,

> *We annoy'd them by Light flanking parties and moved on briskly to the Mill, but were so closely press'd by Col* [James] *Webster's Brigade & Lt Col Tarltons Legion that I found it absolutely necessary to leave a covering party under the Command of Col* [William] *Preston.*[21]

When they reached the creek, Colonel Preston's riflemen were ordered to hold their position and screen the rest of the light corps as they crossed. Captain Joseph Graham with a company of mounted North Carolina militia, witnessed this and recalled,

> [Colonel Williams] *stationed two companies of riflemen, behind trees, one on each side of the road. Thirty poles behind these, as the ground began to turn, he formed a line of militia facing the enemy.*[22]

As Lee's Legion and the rest of the continental light troops safely crossed Reedy Fork Creek and formed on a ridge overlooking the north bank, "*a brisk fire began on Col. Prestons Party which they return'd with great Spirit...*"[23]

---

[21] Ibid.
[22] Graham, 343
[23] Showman, ed., "Col. Otho Williams to General Greene, 7 March, 1781," *The Papers of General Nathanael Greene*, Vol. 7, 407

Captain Graham described the brief stand of the militia riflemen.

> *The day was still cloudy, a light rain falling at times; the air was calm and dense. The riflemen kept up a severe fire, retreating from tree to tree to the flanks of our second [militia] line. When the enemy approached this, a brisk fire commenced on both sides.... They became enveloped in smoke; the fire had lasted but a short time, when the militia were seen running down the hill from under the smoke. The ford was crowded, many passing the watercourse at other places. Some, it was said, were drowned.*[24]

Preston's men fought well but suffered for their obstinance. Cornwallis bragged later that, *"the back mountainmen and some militia suffered considerably, with little loss on our side."*[25] Banastre Tarleton estimated American losses at Weitzel's Mill to be upwards of a hundred killed, wounded, or captured versus only thirty British casualties.[26] American accounts of their losses were much lower. Colonel Williams reported to General Greene that upwards of twenty-five militia were killed or wounded at Weitzel's Mill as well as a handful of continentals.[27]

With the entire British army bearing down on the ford, Colonel Williams was anxious to continue his retreat towards General Greene. Colonel Williams recalled,

---

[24] Graham, 343
[25] Tarleton, "Earl Cornwallis to Lord George Germain, March 17, 1781," 266
[26] Tarleton, 238
[27] Graham, 346-47

> *I waited only 'till Col. Preston cross'd and then order'd the Troops to retire. The Enemy pursued some distance but receiving several severe checks from small covering parties and being cow'd by our Cavalry [Cornwallis] thought proper to Halt. We continued to retire about five miles....* [28]

Lieutenant Colonel Lee and his Legion delivered some of the checks to the British advance referred to by Colonel Williams.

Although the American light corps had performed admirably and casualties were relatively light, significant resentment existed among Colonel Preston's militia riflemen about how they were used in the battle. Many felt they had been inappropriately endangered and sacrificed at Weitzel's Mill to protect the continental light troops and as soon as they got the chance many of the riflemen left the army for home.[29] One Virginia officer reported the dissention among the militia to Governor Thomas Jefferson a few days after the battle.

> *On the late Skirmish...the Riflemen complained that the burden, and heat, of the Day was entirely thrown upon them, and that they were to be made a sacrifice by the Regular Officers to screen their own Troops. Some [of the militia] rejoin'd their Regiments with the main Body and others thought it a plausible excuse for their return home.*[30]

---

[28] Showman, ed., "Col. Williams to General Greene, 7 March, 1781," *The Papers of General Nathanael Greene*, Vol. 7, 407

[29] Graham, 346

[30] Boyd, ed., "Charles Magill to Governor Jefferson, 10 March, 1781," *The Papers of Thomas Jefferson*, Vol. 5, 121

The disgruntlement and departure of the militia temporarily weakened General Greene's army and may have influenced his decision to dissolve the light corps and form two smaller corps of observation under Lieutenant Colonels Lee and Washington. Greene informed Lee of the new arrangement in a letter on March 9th.

> *The light Infantry is dissolvd, and the Army will take upon itself an entire new formation. Col Williams will join the line. And I propose in lieu of the light Infantry two parties of observation, one to be commanded by you, and the other by Lt Col Washington. You are to attend to the enemies movements upon the left wing and Washington upon the right. It is my intention to give Col Washington about 70 or 80 Infantry and between three & four hundred riffleman to act with him. Col Campbell I mean shall join you with about the same number of riflemen, and you and Col Washington either separately or conjunctively as you may agree, to give the enemy all the annoyance in your power, and each to report to Head Quarters.*[31]

Greene confessed that he was uncertain of Cornwallis's intentions and instructed Lee to, "*observe his motions with great attention and give me the earliest information.*"[32]

General Greene also expressed his frustration with the militia but declared that he still planned to pursue Cornwallis once reliable reinforcements arrived.

---

[31] Showman, ed., "General Greene to Colonel Henry Lee Jr., 9 March, 1781," *The Papers of General Nathanael Greene*, Vol. 7, 415
[32] Ibid.

> *I am vexed to my soul with the Militia, they desert us by hundreds nay by thousands. I am now waiting for Gen Caswell* [with North Carolina militia] *reinforcements and the Continental troops* [from Virginia] *to join us.... After which we shall march in pursuit of the enemy as soon as possible, tho our force will not be very respectable.*[33]

Lee informed Greene two days later that Cornwallis had moved to the vicinity of Guilford Court House, nearly twenty miles southwest of the American army. Lee encouraged Greene to, *"get near him"* and asserted, *"I am clear you have a right to fight his Lordship with good reasons to expect success."*[34]

Lee followed his own advice and posted his detachment close to the British. He also maintained constant patrols to guard against a sudden movement by Cornwallis.[35] While General Greene waited along the Haw River for reinforcements to arrive, Cornwallis marched his army a few miles southwest of Guilford Court House. He still hoped to attract Tory support and engage General Greene, but his supply situation was dire so he also made plans to march southeast towards Cross Creek to re-supply and refresh his famished men.[36] These plans were temporarily set aside when Cornwallis learned that General Greene and his army were on the march to Guilford Court House.

---

[33] Ibid.
[34] Showman, ed., "Lt. Col. Lee to General Greene, 11 March, 1781," *The Papers of General Nathanael Greene*, Vol. 7, 427-28
[35] Showman, ed., "Lt. Col. Lee to General Greene, 12 March, 1781," *The Papers of General Nathanael Greene*, Vol. 7, 428
[36] Tarleton, " General Cornwallis to Lord Germain, 17 March, 1781," 267

North Carolina

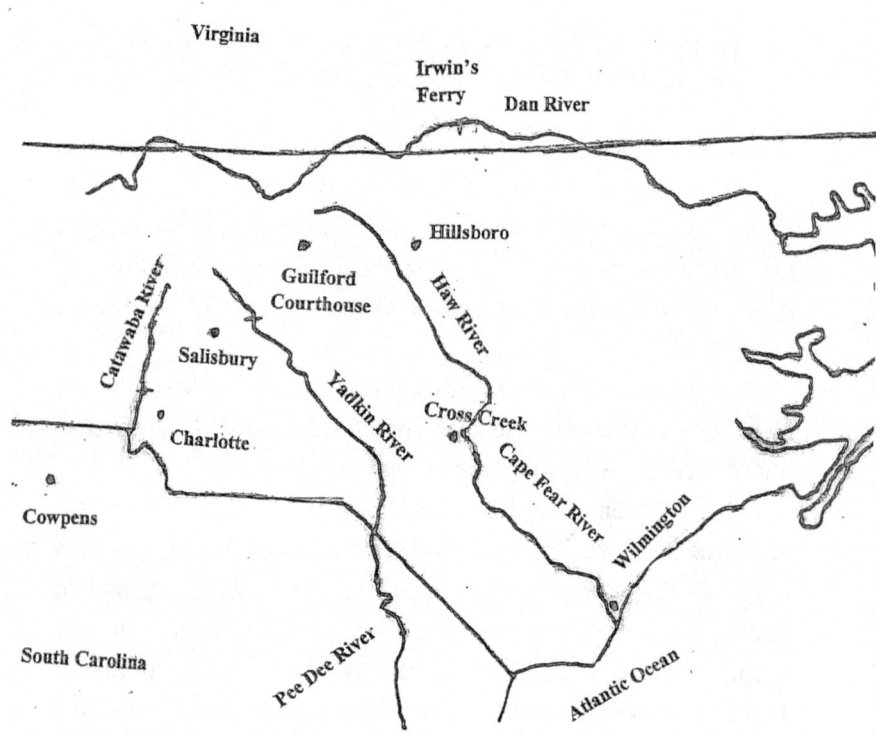

As General Greene had hoped, the arrival of long awaited reinforcements, including 400 Virginia continentals, 1,000 North Carolina militia, and 1,200 Virginia militia, significantly bolstered Greene's force. As a result, General Greene's army surpassed 4,000 men and he decided to act on Lee's advice. He explained his decision to confront Cornwallis in a letter to Congress on March 16th.

> *Finding that our force was much more respectable than it had been, and that there was a much greater probability of its declining than increasing, and that there would be the greatest difficulty in subsisting it long in the field in this exhausted Country, I took the resolution of attacking the Enemy....*[37]

Guilford Courthouse

General Greene's army of 4,400 men bivouacked at Guilford Court House on March 14th, 1781 with plans to attack General Cornwallis the next day.[38] As a precaution against a sudden enemy advance, Greene posted Lieutenant Colonel Washington with part of his corps of observation upon the main road to Salisbury. Lee and his legion, reduced by the hardships of the past two months to about 150 effective dragoons and infantry, were posted three miles west of Guilford Court House along the New Garden Road to guard against a surprise from that direction. They were joined by a portion of Colonel Campbell's riflemen (approximately 60)

---

[37] Showman, ed., "General Greene to Congress, 16 March, 1781," *The Papers of General Nathanael Greene*, Vol. 7, 433
[38] Thomas Baker, *Another Such Victory: The Story of the American Defeat at Guilford Courthouse That Helped Win the War for Independence*, (Ft. Washington, PA: Eastern National, 2005), 40-41

and a company of Virginia continental infantry.[39] Lieutenant Colonel Lee sent a few of his dragoons under Lieutenant James Heard further west to patrol the outskirts of the British camp.

Early in the morning of March 15th, Lieutenant Heard reported that the enemy was on the move. Lee informed General Greene and led his detachment westward to investigate further. Lee's infantry and riflemen were unable to keep pace with his dragoons so they trailed behind at a jog.

Off in the distance, a ragged volley echoed through the early morning. This was soon followed by the arrival of Lieutenant Heard and his detachment, who confirmed that the British army was advancing and that their vanguard, commanded by Banastre Tarleton, was right behind them. Lee ordered his dragoons to retire eastward towards his infantry. As they did so, Tarleton's force caught up to Lee and attacked. The engagement occurred near a section of road that was bordered by steep banks and high fences which limited each side's maneuverability. Confident in the superiority of his cavalry, Lee suddenly swung his dragoons around and charged down the road towards the oncoming enemy. He described the outcome of this bold move in his memoirs.

> *The charge was ordered by Lee, from conviction that he should trample his enemy under foot, if he dared to meet the shock; and thus gain an easy and complete victory. But only the front section of each corps closed, Tarleton sounding a retreat, the moment he discovered [Lee's] column in charge. The whole of the enemy's section was dismounted, and many of the*

---

[39] Babits and Howard, 51

> *horses prostrated; some of the dragoons killed, the rest made prisoners; not a single American soldier or horse [was] injured.*[40]

Lee and his dragoons pursued Tarleton westward towards New Garden Meeting House where they ran into advance troops of the main British army. A volley from the British infantry dismounted a number of Lee's dragoons, including Lee himself. He escaped by taking the mount of one of his troopers, who instantly offered his horse to Lee. The rattled legion commander then ordered his dragoons to retreat.[41] As Lee's cavalry withdrew, his legion infantry and Colonel Campbell's riflemen, no doubt tired and footsore, arrived upon the scene and entered the fray. Lee recalled that,

> *While the cavalry were retiring, the Legion infantry came running up with trailed arms, and opened a well-aimed fire upon the guards, which was followed in a few minutes by a volley from the riflemen under Colonel Campbell, who had taken post on the left of the infantry. The action became very sharp, and was bravely maintained on both sides.*[42]

Concerned that General Cornwallis and the entire British army would arrive at any moment, Lee conducted a successful fighting withdrawal eastward towards Guilford Court House. He had lost approximately 35 men, half of them from his own legion.[43]

---

[40] Lee, 273-74
[41] Babits and Howard, 53
[42] Lee, 274-75
[43] Babits and Howard, 56

Lee's Legion 1781

Battle of Guilford Courthouse

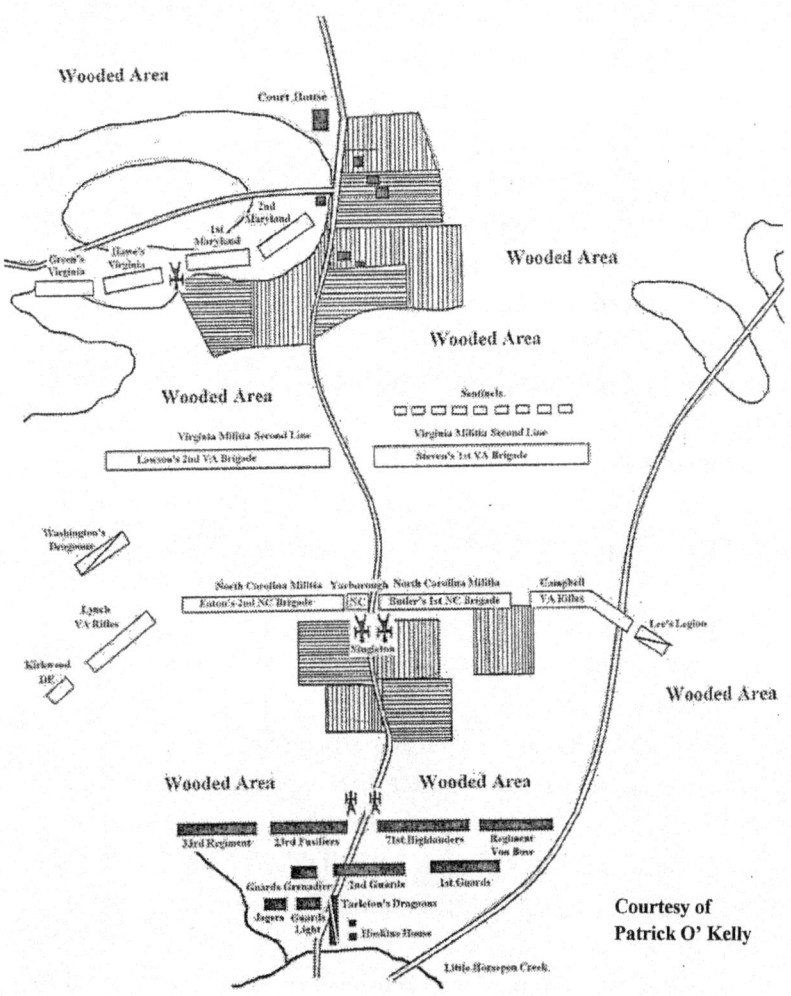

Courtesy of
Patrick O' Kelly

Four miles to the east, General Greene and the rest of the American army heard the battle and prepared to meet the enemy. Greene borrowed the tactics of General Morgan and divided his force into separate lines. Over a thousand North Carolina militia held the first line, which bordered woods and extended along a fence. To their front were open fields over which the British had to march. General Greene asked the militiamen in the first line to fire two good volleys at the British before they retreated to the rear.[44] To anchor this line, Greene placed Lieutenant Colonel Washington's detachment on the right flank and Lieutenant Colonel Lee's force (upon their arrival from the New Garden Meetinghouse fight) on the left. Two six pound cannon were posted in the road, which bisected the militia line down the middle.

General Greene formed a second line of militia about three hundred yards behind the first. These troops, numbering about 1,200, were Virginians under General Robert Lawson and General Edward Stevens. Their line was also split in half by the road and extended into the heavily wooded landscape. The trees and brush provided plenty of cover for the Virginians.

General Stevens was particularly anxious for his men to redeem Virginia's honor after the appalling conduct of the state militia at Camden seven months earlier. Hundreds of Virginians fled the Camden battlefield without firing a shot and General Stevens was determined to prevent a repeat of that humiliation. To insure that his men held their ground, Stevens posted a number of sentinels twenty yards behind the line with instructions to shoot the first man who retreated without orders.[45]

---

[44] Baker, 55
[45] Lee, 277

The third and last American line was posted on a hill adjacent to the road near the courthouse, approximately eight hundred yards behind the Virginia militia. Fourteen hundred continentals from Maryland and Virginia held this position.[46] Most of the Virginian continentals were newly raised recruits with little military experience. They were placed on the right of the line and were commanded by General Isaac Huger. Colonel Otho Williams commanded the more seasoned Maryland troops on the left.

The Battle of Guilford Courthouse commenced around noon when elements of Cornwallis's army crossed Little Horsepen Creek, eight hundred yards in front of General Greene's first line.[47] Greene's artillery opened fire on the British, who responded with their own cannon fire. While the artillery crews blasted away, General Cornwallis crossed the creek with the rest of his army and deployed his troops along a thousand yard front.[48]

Around 1:00 p.m., approximately 2,000 British and German soldiers advanced over open ground towards General Greene's first line. The North Carolinians anxiously waited for the enemy to advance within range of their smoothbore muskets, preferably one hundred yards or less. The tension proved unbearable, and they fired a ragged volley at the enemy at 140 yards.[49]

---

[46] Baker, 52
[47] Ibid., 56
[48] Ibid., 59
[49] Showman, ed., "General Greene to Samuel Huntington, 16 March, 1781," *The Papers of General Nathanael Greene*, Vol. 7, 434 see also note 9 on pages 437-439

The British continued their advance, sparking panic in the center of the North Carolina line. Many of the militia dropped their gear and fled to the rear. The troops on the flanks held, however, and leveled their muskets and rifles for another volley. Roger Lamb, a British soldier participating in the attack against the American right wing, recalled,

> *When [we] arrived within forty yards of the enemy's line, it was perceived that their whole force had their arms presented, and resting on a rail fence...They were taking aim with the nicest precision.... At this awful period a general pause took place; both parties surveyed each other for the moment with the most anxious suspense....Colonel Webster rode forward in the front of the 23d regiment, and said..."Come on, my brave Fuzileers!" This operated like an inspiring voice, they rushed forward amidst the enemy's fire; dreadful was the havoc on both sides.*[50]

The American volley found its mark, and scores of Cornwallis's men fell. Captain Dugald Stuart of the 71st Highland Regiment noted that, *"one half of the Highlanders dropt on the spot."* [51] Sir Thomas Saumarez of the Fusiliers remembered it as a, *"most galling and destructive fire."*[52] The volley, though deadly, did not stop the British advance. Banastre Tarleton recalled that *"The King's troops threw in their fire, and charged rapidly with their bayonets."*[53] The

---

[50] Roger Lamb, *An Original and Authentic Journal of Occurrences During the Late American War*, (Dublin: Wilkinson & Courtney, 1809), 361
[51] Buchanan, 375
[52] Ibid.
[53] Tarleton, 273

remaining Americans in the first line were forced back as the British swept over the rail fence and continued towards General Greene's next line, three hundred yards away.

General Greene's plan called for Lieutenant Colonel Washington and Lieutenant Colonel Lee to withdraw to the flanks of the third line, but Lee was forced to veer to the southeast, away from the American position. He soon found himself engaged in a completely separate struggle. His legion and Colonel Campbell's Virginia riflemen, along with 120 North Carolina militia who were swept along with Lee, faced a battalion of British Guards in a pitched battle in the woods. The fight quickly became confused as the heavily wooded terrain broke up units. Pockets of men fought from tree to tree in all directions. General Cornwallis described this part of the battle.

> *The excessive thickness of the woods rendered our bayonets of little use, and enabled the broken enemy to make frequent stands, with an irregular fire, which occasioned some loss...particularly on our right, where the 1st battalion* [of Guards] *and regiment of Bose* [Germans] *were warmly engaged in front, flank, and rear...with part of their extremity of their left wing....*[54]

Lee and his men acquitted themselves well against the British Guards and made them pay dearly for the ground they took. This was especially true when the left end of General Stevens's militia swung into the flank and rear of the Guards as the heavily engaged Guards passed by Stevens's line

---

[54] Tarleton, "Lord Cornwallis to Lord George Germain, 17 March, 1781," 305-307

without noticing them. Samuel Houston was stationed on the left flank of Stevens's line and recalled,

> *We fired on their flank, and that brought down many of them...We pursued them...to the top of a hill, where they stood, and we retreated from them back to where we formed.*[55]

Charles Stedman, a British participant of the battle, described the impact on the Guards:

> *At one period of the action the first battalion of the guards was completely broken. It had suffered greatly in ascending a woody height...[and] no sooner was it done, than another line of the Americans presented itself to view, extending far beyond the right flank of the guards, and inclining towards their flank, so as almost to encompass them.... The enemy's fire...being poured in not only on the front but flank of the battalion, completed its confusion and disorder, and notwithstanding every exertion made by the remaining officers, it was at last entirely broken.*[56]

The British Guards were spared from complete destruction by the timely arrival of a Hessian regiment. Their presence provided the Guards with an opportunity to reform their ranks. The frantic battle continued:

---

[55] Baker, 64
[56] C. Stedman, *The History of the Origin, Progress, and Termination of the American War*, Vol. 2 (London: 1794), 341-342

> *No sooner had the guards and Hessians defeated the enemy in front, than they found it necessary to return and attack another body of them that appeared in the rear; and in this manner they were obliged to traverse the same ground in various directions....*[57]

The fighting on his right eventually attracted General Cornwallis's attention. He dispatched Lieutenant Colonel Tarleton and his Legion to investigate. They arrived at the end of the conflict and found, "*a few hardy riflemen...lurking behind trees.*"[58] These were Colonel Campbell's riflemen, who moments earlier were abandoned by Lieutenant Colonel Lee's abrupt withdrawal to the third American line. Lee justified his sudden departure from the flank fight with the claim in his memoirs that the fighting there had largely wound down, but the departure of Lee's Legion left Colonel Campbell's riflemen defenseless against Tarleton's saber wielding dragoons and many were cut down. Campbell was furious with Lee and left the army days after the battle. Most of his riflemen went with him.[59]

While these events unfolded on the far left on the second American line, over a thousand Virginia militiamen to Lee's right were engaged in their own desperate struggle to stop the British advance. The main British line approached Greene's second position with some difficulty; the dense woods broke up their formation.[60] As a result, British units attacked the

---

[57] Ibid., 343
[58] Ibid.
[59] Lyman C. Draper, *King's Mountain and Its Heroes: History of the Battle of King's Mountain*, (TN : Over mountain Press, 1996), 394 (Originally published in 1881)
[60] Tarleton, 273-274

Virginians in piecemeal fashion rather than in one continuous line. This reduced the shock value of Cornwallis's assault and helped the Virginians stand their ground. British historian Charles Stedman noted that,

> *The second line of the enemy made a much braver and stouter resistance than the first. Posted in the woods, and covering themselves with trees, they kept up for a considerable time a galling fire, which did great execution.*[61]

Henry Lee recounted in his memoirs that the Virginians fought so well that they forced General Cornwallis to commit all of his reserves:

> *Noble was the stand of the Virginia militia; Stevens and Lawson, with their faithful brigades...*[caused] *every corps of the British army, except the cavalry,* [to be] *brought into battle, and many of them suffered severely.*[62]

Major St. George Tucker, an officer in General Lawson's militia brigade, was less complimentary of his men's effort. When word spread that the enemy had gained their right flank, most of Tucker's men fled to the rear. He rallied about seventy of them and engaged in an *"irregular kind of skirmishing with the British, and were once successful enough to drive a party for a very small distance."*[63] Tucker added that the men who stayed with him fired up to twenty rounds in

---

[61] Stedman, 339
[62] Lee, 278-279
[63] Commager and Morris, "Major St. George Tucker to his Wife, 18 March, 1781," 1166

the conflict, a significant number compared to the performance of the North Carolinians in the first line. Nonetheless, because most of his unit fled early in the engagement, Tucker was disillusioned by their conduct.

General Greene, on the other hand, was pleased with the Virginia militia's conduct and noted the intensity of their effort in a letter to Congress. He declared, *"The Virginia Militia gave the Enemy a warm reception and kept up a heavy fire for a long time...."*[64] He similarly informed Governor Thomas Jefferson that the British *"Were very much galled by an incessant fire* [from the Virginia militia]."[65]

Most observers agreed that the Virginians fought well. Eventually, however, with General Stevens wounded in the thigh and British troops pressing their front and flanks, the Virginians were forced to retreat. They scurried back towards the last American line, pressed closely by the enemy.

Just as the British did at the second line, Cornwallis's troops arrived at Greene's third line in sections. The first to arrive was Lieutenant Colonel James Webster and the left wing of the British line. Instead of waiting for the right wing of the army to catch up, Webster unwisely led his brigade forward against Greene's continentals. The British charged into a cleared ravine and then up a hillside towards the Americans. The Virginia and Maryland troops coolly held their fire until the British were at point blank range and then blasted them with a murderous volley. Webster's brigade halted in its tracks, their commander mortally wounded in the

---

[64] Showman, ed. "General Greene to Samuel Huntington, 16 March, 1781," *The Papers of General Nathanael Greene,* Vol. 7, 435

[65] Boyd, ed., "General Greene to Thomas Jefferson, 16 March, 1781," *The Papers of Thomas Jefferson,* Vol. 5, 156

leg, and retreated to the far wood line to await reinforcements.⁶⁶

Additional British units arrived on Webster's right. British Lieutenant Colonel James Stuart led the 2ⁿᵈ Battalion of Guards against the left of the continental line, which was held by the 2ⁿᵈ Maryland Regiment. The Marylanders fired a weak volley at the oncoming British and then inexplicably broke ranks and fled, leaving two cannon and the left flank of General Greene's third line dangerously exposed. General Greene reacted quickly to prevent a complete rout of his army and ordered a retreat, but before the order was executed, Colonel John Gunby, commander of the 1ˢᵗ Maryland Regiment, *"wheeled to his left upon Stuart, who was pursuing the flying second regiment."*⁶⁷ One observer noted that

> *This conflict between the brigade of guards and the first regiment of Marylanders was most terrific, for they fired at the same instant, and they appeared so near that the blazes from the muzzles of the guns seemed to meet.*⁶⁸

Gunby's men followed their volley with a bayonet charge. At nearly the same moment, Colonel William Washington and his dragoons swept down on the British. The Guards, who up to that point had kept their order, *"were driven back with great slaughter"* by the cavalry's sabers and the Marylanders' bayonets.⁶⁹ One Virginian horseman in particular, Peter Francisco, a giant of a man at six and a half feet tall,

---

⁶⁶ Lee, 279
⁶⁷ Ibid.
⁶⁸ Baker, 73
⁶⁹ Lamb, 351

reportedly struck down numerous guardsmen before he was bayoneted in the leg.[70] The Maryland troops also inflicted great loss on the British. Lieutenant Colonel John Eager Howard, who took command of the 1st Maryland when Colonel Gunby was pinned under his horse, proudly recalled:

> *My men followed quickly, and we pressed through the guards, many of whom had been knocked down by the horse without being much hurt. We took some prisoners, and the whole were in our power.*[71]

The bold charge of the 1st Maryland and William Washington's cavalry briefly steadied the American left flank, but British artillery fire prevented the Americans from capitalizing on their success. As a result, Washington's cavalry and the Maryland continentals disengaged and joined the rest of the American army in an orderly retreat from Guilford Courthouse. The British were too battered to vigorously pursue, so the battle came to a close.

Although Guilford Courthouse ended as another American defeat in the south, the victors had little to celebrate. Cornwallis lost over one quarter of his effective force, including many key officers. His army was exhausted and weak and in desperate need of rest and provisions. Three days after the bloody fight the British army limped towards Wilmington where they hoped to be re-supplied by the navy.

The departure of most of the American militia soon after the battle and a severe shortage of ammunition prevented General Greene from aggressively pursuing Cornwallis. Instead, he ordered Lieutenant Colonel Lee and his legion to

---

[70] Baker, 74
[71] Ibid., 75

pursue and harass the British on their march southward and slow them down in hopes that Greene might receive reinforcements and ammunition in time to catch Cornwallis and engage him again.

Chapter Nine

"Our poor Fellows are worne out with fatigue"

General Greene and his battered army followed Lee a few days after Lee's departure. Despite the absence of most of his militia (who returned to their homes after the battle) Greene informed Lee that he intended to attack Cornwallis. He urged Lee to, *"push their rear all you can,"* to slow Cornwallis down.[1] Lee was unable to comply without risking his men and it was soon evident that long before Greene could catch him, Cornwallis would reach Cross Creek (modern day Fayetteville) where he hoped to be re-supplied.

Travelling in the wake of the famished British army proved to be a challenge for the Americans. Despite the fact that Greene's army had shrunk by half in the days following Guilford Court House it was still extremely difficult for Greene's men to gather enough provisions in areas that the British army had passed through. Short on men and supplies, General Greene was eventually forced to halt his pursuit at Ramsay's Mill.[2] In frustration he settled upon a new strategy

---

[1] Showman, ed., "General Greene to Lt. Col. Lee, 22 March, 1781," *The Papers of General Nathanael Greene,* Vol. 7, 461
[2] Showman, ed., "General Greene to Col. S. Drayton, 28 March, 1781," *The Papers of General Nathanael Greene,* Vol. 7, 475

and shared it with General Washington in a letter on March 29th.

> *In this critical and distressing situation I am determined to carry the War immediately into South Carolina. The Enemy will be obliged to follow us or give up their posts in that State.*"[3]

General Greene had resolved to lead his army back into South Carolina and force General Cornwallis to make a choice, pursue Greene and protect British control of South Carolina, or abandon South Carolina and march his army into Virginia.

General Greene spent a few more days at Ramsay's Mill gathering what little provision he could before he commenced his march southward towards Camden, South Carolina on April 6th. Lieutenant Colonel Lee and his legion, reinforced by a detachment of Maryland continentals, also marched south, but not towards Camden. Lee was ordered to reunite with General Francis Marion and attack several British outposts along the Santee River.[4] Their first target was Fort Watson.

---

[3] Showman, ed., "General Greene to General Washington, 29 March, 1781," *The Papers of General Nathanael Greene*, Vol. 7, 481

[4] Dennis M. Conrad, ed., "General Francis Marion to General Nathanael Greene, 21 April, 1781," *The Papers of Nathanael Greene*, Vol. 8, 129 See also: Lee, 333

British Outposts in South Carolina

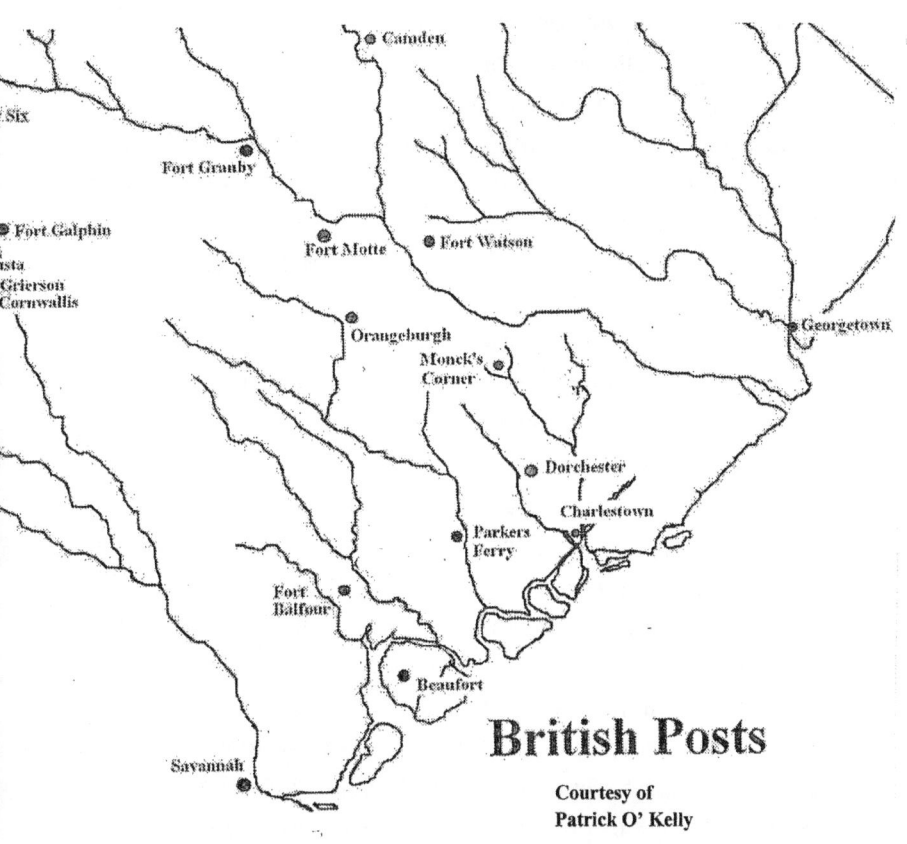

British Posts

Courtesy of
Patrick O' Kelly

Fort Watson

Fort Watson was one of a string of fortified outposts the British had built in central South Carolina in 1780-81 to secure their communication and supply lines between Charleston, and their main outposts at Georgetown, Camden, and Ninety Six. The fort was securely situated atop an Indian burial mound on the east bank of the Santee River. General Marion informed General Greene in a letter that the fort sat on top of a, *"Small hill,* [with a] *stockade...*[and] *three rows of Abbatis around it, no trees near enough to cover our men."*[5]

Although the mixed British and Tory garrison of 114 men was significantly outnumbered, Lieutenant Colonel Lee and General Marion ruled out a frontal assault.[6] A siege of the fort was also dismissed because of the absence of cannon or entrenching tools. Instead, Lee and Marion sought to surround the fort and sever the garrison's access to fresh water from a nearby lake. This move was effectively countered by the garrison when they successfully dug a well.[7]

With mounting fears that British reinforcements might soon arrive to relieve the fort, Lee and Marion approved a novel idea from Major Hezekiah Maham of Marion's militia brigade. Maham proposed the construction of a 40 foot wooden tower to serve as a firing platform for some riflemen. Maham's tower would allow the marksmen to shoot over the garrison's walls from a relatively sheltered position and target the interior of the fort. According to Lee, Major Maham,

---

[5] Conrad, ed., "General Marion to General Greene, 23 April, 1781," *The Papers of General Nathanael Greene*, Vol. 8, 139
[6] Ibid., 141
[7] Ibid., 139

> *Proposed to cut down a number of suitable trees in the nearest wood, and with them to erect a large strong oblong pen, to be covered on the top with a floor of logs, and protected on the side opposite the fort with a breastwork of light timber....*[8]

Construction of the tower took a few days. Lee described its effect in his memoirs.

> *A party of riflemen, being ready, took post in the Maham tower the moment it was completed; and a detachment of musketry, under cover of the riflemen, moved to make a lodgment in the enemy's ditch, supported by the Legion infantry with fixed bayonets in the nets. Such was the effect of the fire from the riflemen, having thorough command of every part of the fort, from the relative supereminence of the tower, that every attempt to resist the lodgment was crushed.*[9]

The impact of the American rifle fire from the tower upon the garrison's morale was dramatic. According to the journal of a British officer in the fort, the American rifle fire prompted, "*the Cowardly & Mutinous behavior of A Majority of the Men, [who] grounded their Arms & refused to defend the Post any longer.*"[10] Lieutenant James McKay, the garrison's commander, had no choice but to surrender on April 23rd.

---

[8] Lee, 332
[9] Ibid.
[10] Conrad, ed., "Journal of British Officer at Fort Watson, Note 1," *The Papers of General Nathanael Greene*, Vol. 8, 141

While Lee and Marion successfully overcame British resistance at Fort Watson, General Greene and his army (reduced to less than 1,500 men) encamped on the outskirts of Camden, South Carolina.[11] Camden was a key British outpost in central South Carolina that was well fortified and defended. Twenty-six year old Lord Francis Rawdon commanded the 900 man British garrison at Camden.[12] Despite the close proximity of the two armies, nearly a week passed before an engagement ensued. General Greene explained the delay in a letter to Congress.

> *On our arrival at Camden we took post at Log Town, about half a mile in front of their Works, which upon reconnoitering were found to be much stronger than had been represented, and the garrison much larger.... Our force was too small either to invest the Town or storm the Works, which obliged us to take a position at a little distance from it.*[13]

Hoping to draw Lord Rawdon from his fortified position, General Greene posted his troops on a hill a mile north of town. He soon got his wish.

---

[11] Lee, 333

[12] Conrad, ed., "General Greene to the President of Congress, 27 April, 1781," *The Papers of General Nathanael Greene*, Vol. 8, note 1, 157

[13] Conrad, ed., "General Greene to the President of Congress, 22 April, 1781," *The Papers of General Nathanael Greene*, Vol. 8, 131

Hobkirk's Hill

On the morning of April 25th Lord Rawdon led a force of over 900 troops, including armed, "*Musicians...Drummers & in short everyone that could carry a Firelock,*" against General Greene's troops on Hobkirk's Hill.[14] Rawdon informed General Cornwallis that the impetus to attack Greene's larger force were reports that Greene's artillery was a day's march in his rear and that reinforcements under Lee, Marion, and Sumter were on the way to join Greene.[15] Rawdon hoped to defeat Greene before these reinforcements made Rawdon's defense of Camden untenable.

In an effort to surprise the Americans, Rawdon led his force on a circuitous route through thick woods to Hobkirk's Hill. Fortunately for General Greene, Rawdon's approach was discovered by his pickets whose "*warm reception,*" of the British alerted General Greene and his army of the impending attack.[16]

Greene posted his three 6 pound artillery pieces, which had actually arrived that morning, in the center of his line and placed his two regiments of Maryland continentals to the left of the cannon and his two regiments of Virginia continentals to the right. A small force of 250 militia was held in reserve in the rear.

Confident that he outnumbered the enemy and that he could outflank them, Greene ordered his men to advance down the hill while his cavalry under Lieutenant Colonel William

---

[14] Conrad, ed., "Lord Rawdon to General Cornwallis, 25-26 April, 1781, Note 1," *The Papers of General Nathanael Greene*, Vol. 8, 157
[15] Ibid.
[16] Conrad, ed., "General Greene to Congress, 27 April, 1781," *The Papers of General Nathanael Greene*, Vol. 8, 155

Washington circled around the enemy to strike their rear.[17] Greene explained his actions to Congress in a letter the day after the battle.

> *As the Enemy were found to be advancing only with a small front, Lieut Col. Ford with the 2nd Maryland Regiment had orders to advance and flank them upon the left, Lieut Col. Campbell* [commanding a regiment of Virginia continentals] *had orders to do the like upon the right. Col. Gunby with the first Maryland Regiment, and Lieut Col. Haws with the second Virginia Regiment, had orders to advance down the Hill and charge them in front. Lieut Col. Washington had orders to turn the Enemies right flank and charge them in the rear. The whole line was soon engaged in close firing, and the Artillery under Col. Harrison playing on their front. The Enemy were staggered in all quarters, and upon the left were retiring while our Troops continued to advance....*[18]

William Seymour, a sergeant with the Delaware continentals, was with Greene's advance troops when the attack began and observed that the American cannon fire,

> *Put the enemy in great confusion, having killed and dangerously wounded great numbers.*[19]

---

[17] Ibid.
[18] Ibid.
[19] Conrad, ed., "General Greene to the President of Congress, 27 April, 1781, Note 3," *The Papers of General Nathanael Greene*, Vol. 8, 158

General Greene, who was with Lieutenant Colonel Hawes's 2nd Virginia Regiment near the center of the American line, also noted the initial success of the American attack. He recalled that the artillery, *"was doing great execution,"* and that Hawes's Virginians were advancing, *"in tolerable order, within forty yards of the Enimy and they in confusion in front."*[20]

General Greene's double envelopment appeared to be working when suddenly the attack unraveled. Disregarding orders not to fire but to advance on the enemy with bayonets only, the Maryland troops on the left paused to fire a volley and while doing so, the commander of the 1st Maryland Regiment's far right company, Captain William Beatty, was killed, throwing his company into confusion.[21] At almost the same moment, Lieutenant Colonel Ford of the 2nd Maryland Regiment (on the left of Gunby's regiment) was seriously wounded. The loss of these officers produced confusion and uncertainty among their troops. General Greene noted:

> *Unfortunately two Companies of the right of the first Maryland Regiment got a little disordered, and unhappily Col Gunby gave an order for the rest of the Regiment then advancing to take a new position in the rear, where the two Companies were rallying. This impressed the whole Regiment with an Idea of a retreat, and communicated itself to the 2nd Regiment*

---

[20] Conrad, ed., "General Greene to Lt. Col. Lee, 28 April, 1781," *The Papers of General Nathanael Greene*, Vol. 8, 168

[21] Conrad, ed., "General Greene's Orders, 2 May, 1781," *The Papers of General Nathanael Greene*, Vol. 8, 187

which immediately followed the first on their retiring.[22]

Lieutenant Colonel Gunby's decision to retreat and reform his regiment upon his two disordered companies proved to be a crucial mistake. The British boldly advanced into the gap in the American line that was formed by Gunby's withdrawal and threatened Greene's artillery, which was forced to the rear. Rawdon's troops also turned the left flank of Lieutenant Colonel Hawes 2nd Virginia. Greene noted that

> *The second Virginia Regiment having advanced some distance down the Hill, and the Maryland line being gone the Enemy immediately turned their flank, while they were engaged in front. Lieut Col. Campbell's* [Virginia] *Regiment had got into some disorder and fallen back a little, this obligated me to order Lieut Col. Haws to retire.*[23]

The withdrawal of General Greene's last regiment signaled the end of the battle, and although General Greene was bitterly disappointed with the result, he correctly observed that the British paid a high price for their victory. Over one quarter of Rawdon's force, 258 men, were casualties. The American's suffered similar losses, with 270 men killed, wounded, captured or missing.[24]

---

[22] Conrad, ed., "Greene to Congress, 27 April, 1781," *The Papers of General Nathanael Greene*, Vol. 8, 156

[23] Ibid.

[24] Conrad, ed., "General Greene to Congress, 27 April, 1781, note 3," *The Papers of General Nathanael Greene*, Vol. 8, 158

General Greene's army withdrew northward a few miles, screened by Lieutenant Colonel Hawes's Virginians and Lieutenant Colonel Washington's cavalry. Rawdon's pursuit was limited and ineffectual and he returned to Camden before nightfall. Although General Greene had lost the battle, his army remained intact and a threat to Rawdon.

Thirty miles away, Lieutenant Colonel Lee and General Marion, unaware of events at Camden, impatiently waited in the High Hills of the Santee for news and instructions from General Greene. Lee was eager to strike at more British outposts along the Santee River and requested a field piece and reinforcements.[25] Marion complained that he had received no reply to his last three letters and that he feared Greene had retreated from Camden, leaving Lee and himself in a dangerous situation.[26] Both commanders learned of the turn of events at Hobkirk's Hill on April 28th. Lee, as he did in earlier letters, requested permission to pass to the west side of the Santee River with the artillery piece Greene was sending. Lee wanted to, *"pursue the conquest of every post & detachment in that country,"* and believed that the fall of these outposts would force the British to abandon Camden by severing the garrison's supply line.[27] Lee asserted that the only other way to capture Camden was to storm it, something none of the Americans relished.

---

[25] Conrad, ed., "Lt. Col. Lee to General Greene, 27 April, 1781," *The Papers of General Nathanael Greene*, Vol. 8, 162
[26] Conrad, ed., "General Marion to General Greene, 27 April, 1781," *The Papers of General Nathanael Greene*, Vol. 8, 163
[27] Conrad, ed., "Lt. Col. Lee to General Greene, 28 April, 1781," *The Papers of General Nathanael Greene*, Vol. 8, 171

General Greene's reply to Lee conveyed a sense of annoyance and frustration at his eager yet outspoken subordinate.

> *You write as if you thought I had an army of fifty thousand Men. Surely you cannot be unacquainted with our actual situation. I have run every risqué to favor your operations more perhaps than I ought.... I am strongly impressed with the necessity of pushing our operations on the west side of Santee as you can be, but the means are wanting. We want reinforcements. You want detachments; and if you and General Marion separate you will both be exposed.*[28]

Despite the critical tone of Greene's reply, the general informed Lee that he had already acquiesced to his request, but he urged Lee to be cautious.

> *In my letter to General Marion last Evening, I desird him either to detach you, or cross the Santee with you.... I beg you not to think of running great hazards, our situation will not warrant it.*[29]

Despite General Greene's cautious approval, Lee's plans were temporarily disrupted when he learned that a force of approximately 500 Tories under Colonel John Watson was marching from Georgetown to Camden. Lee and Marion attempted to intercept Watson but failed. The arrival of

---

[28] Conrad, ed., "General Greene to Lt. Col. Lee, 29 April, 1781," *The Papers of General Nathanael Greene*, Vol. 8, 172-73
[29] Ibid.

Watson's troops significantly bolstered Lord Rawdon's garrison at Camden. Rawdon attempted to attack General Greene the day after Watson's arrival, but the Americans had wisely redeployed to a stronger position further from town, so Rawdon returned to Camden disappointed. Two days later, with his supply lines threatened by Lee and Marion, Rawdon abandoned Camden and marched south towards Charleston. Camden, a key British outpost from which control of central South Carolina was exercised, was no longer in enemy hands.

Fort Motte & Fort Granby

A few days before Rawdon abandoned Camden, Lieutenant Colonel Lee and General Marion commenced a siege on Fort Motte, an important British depot along the Camden supply line located where the Congaree and Wateree Rivers joined to form the Santee River. The fort was situated on a commanding hill and was built around a large mansion belonging to Mrs. Rebecca Motte, the widow of an ardent patriot. A deep trench and high earthen wall created a strong position defended by 150 men and a small detachment of cavalry.[30]

Lieutenant Colonel Lee had posted his legion on a nearby hill to the north of the fort while General Marion occupied the ground to the east. The Americans shelled the fort, which lacked any cannon to reply, but the stubborn garrison refused to surrender and remained hopeful that Lord Rawdon would soon arrive to relieve them. By May 12th, two days after Camden was abandoned, concern that Rawdon might indeed

---

[30] Lee, 345

arrive prompted Lee and Marion to adopt an unorthodox tactic to seize the fort.

Mrs. Motte's mansion had been incorporated into the fort's defenses. If the building were burned the garrison might be compelled to surrender. Lee and Marion decided to torch the mansion with flaming arrows and informed Mrs. Motte of their decision. Lee recalled in his memoirs that,

> *With a smile of complacency this exemplary lady listened to the embarrassed officer and gave instant relief to his agitated feeling by declaring, that she was gratified with the opportunity of contributing to the good of her country....*[31]

Mrs. Motte even supplied the bow and arrows that her deceased husband had imported from India.[32]

After the garrison rejected one more appeal to surrender, the Americans carried out their plan. Flaming arrows successfully lodged in the roof of the mansion and ignited the house. British soldiers attempted to extinguish the flames, but American artillery and small arms fire drove them off the roof. The British commander realized his dire situation and surrendered the fort.

A day after the capture of Fort Motte, Lee received orders to march his legion to Fort Granby (in present day Columbia, SC) and demand the surrender of its garrison.[33] Although Fort Granby was strongly situated and defended by over 300 Tories and provincial troops under Major Andrew Maxwell, Lee

---

[31] Ibid., 347
[32] Ibid.
[33] Conrad, ed., "General Greene to Lt. Col. Lee, 13 May, 1781," *The Papers of General Nathanael Greene*, Vol. 8, 249

successfully gained its surrender without the loss of a single man.³⁴ He explained his actions to General Greene.

> *The troops under my command after a rapid march from the post at Motte's reached this place on the evening of* [May] *14th. The enemy's position was reconnoitered, labourers collected, & the line of approaches determined on. During the night a battery was thrown up in point blank shot of the fort.... Early in the morning of the 15th Capt. Finley commenced a cannonade, at the same time the Legion Infantry & Capt. Oldham's detachment* [of Maryland continentals] *took a position of support. Matters being thus arranged, agreeable to my instructions I sent the summons...*³⁵

Eager to secure the fort before British reinforcements arrived and aware that Major Maxwell had a reputation for caring more about the spoils of war he and his men had acquired than any "military laurels" they had earned in battle, Lee offered to let the Tories keep all of their private baggage, no questions asked. In the face of Lee's strong show of force and under such generous terms, Major Maxwell complied with Lee's summons and surrendered the fort. The garrison and their baggage were escorted, for their own protection, to Charleston as paroled prisoners of war to await exchange.

The day after Fort Granby's surrender, General Greene ordered Lee and his legion to march southwestward to join General Andrew Pickens in Augusta, Georgia. Lee pushed his

---

[34] Lee, 352
[35] Conrad, ed., "Lt. Col. Lee to General Greene, 15 May, 1781," *The Papers of General Nathanael Greene*, Vol. 8, 263-64

men hard, often riding two men to a horse, and covered the seventy-five miles in two days.[36] Upon learning of Lee's rapid march, General Greene wrote approvingly to his energetic Lieutenant Colonel.

> Your early arrival at Augusta astonishes me. For rapid marches you exceed Lord Cornwallis and every body else. I wish you may not have injured your troops.[37]

Fort Galphin

At Augusta, Lee learned that the annual royal present to the Cherokee and Creek Indians was stored at Fort Galphin, twelve miles below Augusta. Despite the need to rest his tired force, Lee seized the opportunity to capture the fort, its small garrison and the large store of supplies meant for the Indians. He pushed on to Fort Galphin with those of his men who could continue.

Aware of the contempt the British had for the American militia, Lee used a detachment of mounted militia that accompanied him as bait to draw Fort Galphin's garrison out. He had the militia dismount and advance upon the fort on foot to attract the garrison's attention. Lee expected the British troops inside the fort to rush out in pursuit of the militia, at which point part of Lee's infantry would rush the fort's entrance from the nearby woods while Lee's cavalry and the rest of his infantry rescued the militia. Lee's plan worked flawlessly and he described it years later in his memoirs.

---

[36] Lee, 353-54
[37] Conrad, ed., "General Greene to Lt. Col. Lee, 22 May, 1781," *The Papers of General Nathanael Greene*, Vol. 8, 291

> *The major part of the garrison, as had been expected, ran to arms on sight of the militia, and, leaving the fort, pursued them.... Captain Rudolph was ordered to rush upon the fort, while [some legion infantry], supported by a troop of dragoons...shielded the militia. Rudolph had no difficulty in possessing himself of the fort, little opposition being attempted, and that opposition being instantly crushed. We lost one man from the heat of the weather; the enemy only three or four in the skirmish.*[38]

Yet another British outpost, with over 120 men, had fallen to Lee at the loss of just one man. Additionally, the supply of captured powder and shot from the fort significantly replenished American stocks.[39]

Siege of Augusta

After a few hours of rest, Lee pushed his force across the Savannah River and marched toward Augusta. Attempts to convince the garrisons of the two forts at Augusta (Fort Cornwallis and Fort Grierson) to surrender were spurned by Lieutenant Colonel Thomas Brown, the Loyalist commander. Lee joined General Andrew Pickens and Colonel Elijah Clarke of Georgia, and the three commanders devised a plan to capture the two forts by storming Fort Grierson, the weaker of the two enemy positions. Defended by only 80 Tories and separated from the much larger garrison of Fort Cornwallis by half a mile, the Americans were confident that Fort Grierson

---

[38] Lee, 355
[39] Henry Lumpkin, *From Savannah to Yorktown: The American Revolution in the South*, (toExcel, 1987), 187 and Lee, 355

could be stormed. What remained to be seen was whether Lieutenant Colonel Brown would risk any of his 350 men in Fort Cornwallis to assist Fort Grierson.

American operations against Fort Greirson commenced on the morning of May 24th with the militia of General Pickens and Colonel Clarke attacking the fort from three directions. While this occurred, Lieutenant Colonel Lee posted his legion, with a six pound cannon, between Fort Grierson and Fort Cornwallis to block Lieutenant Colonel Brown from aiding Fort Grierson. Lee recounted in his memoirs that,

> *Browne,* [at Fort Cornwallis] *who drawing his garrison out of his lines, accompanied by two field-pieces, advanced with the appearance of risking battle to save Grierson.... This forward movement soon ceased. Browne, not deeming it prudent, under existing circumstances, to presevere in its attempt, confined his interposition to a cannonade, which was returned by Lee, with very little effect on either side.*[40]

Lee's successful obstruction of Lieutenant Colonel Brown and his troops prevented help from reaching Fort Grierson and the beleaguered garrison was easily overwhelmed by the American militia. General Pickens informed General Greene that 30 Tories were killed and 40 captured in the assault, leaving just a handful, including Lieutenant Colonel Grierson, who escaped along the riverbank to Fort Cornwallis.[41]

---

[40] Lee, 357
[41] Conrad, ed., "General Pickens to Lt. Col. Lee, 25 May, 1781," *The Papers of General Nathanael Greene,* Vol. 8, 311

Although the capture of Fort Grierson went smoothly and came at little cost for the Americans, the same was not true for their efforts against Fort Cornwallis. Lieutenant Colonel Brown and his 350 men were determined to hold the fort, and their strong works insured that an American frontal assault like that upon Fort Grierson would be extremely costly. As a result, Lee, Pickens, and Clarke commenced a siege of Fort Cornwallis.

Days of back breaking work for both sides passed as the Americans steadily advanced their trench and earthworks towards the fort and the British strengthened their own works against the inevitable American assault. In late May, Lieutenant Colonel Brown adopted more forceful measures to resist the Americans and interrupt their progress. Under the cover of darkness he launched nighttime sorties that temporarily drove the Americans from their lines. Brutal hand-to-hand combat was waged in the American trenches as the militia and Lee's Legion struggled to hold and then re-take their own works.

It was about this time that the Americans decided to employ a tactic used by Lee and Marion at Fort Watson, namely, a wooden tower to serve as a firing platform. Unlike the tower at Fort Watson, however, this one would be constructed to mount a six pound cannon.

Using an old wooden house outside of Fort Cornwallis as cover, construction of the tower began on the evening of May 30th. Lee recounted that

> *In the course of the night and ensuing day we had brought our tower nearly on a level with the enemy's parapet, and began to fill its body with fascines,*

> *earth, stone, brick, and every other convenient rubbish, to give solidity and strength to the structure.*[42]

Lieutenant Colonel Brown recognized the danger the tower presented to his men and launched a furious assault to destroy it, but his men were driven back at the point of bayonets by the Americans. Lee recalled that, "*Upon this occasion the loss on both sides exceeded all which had occurred during the siege.*"[43]

Brown next tried to demolish the tower with his artillery, but this also failed and on June 2nd the Americans began raking Fort Cornwallis with cannon fire from the tower. They soon dismounted the British cannon. In desperation Colonel Brown sent a sergeant, posing as a deserter, to the American lines to try to burn the tower, but Lieutenant Colonel Lee grew suspicious of the man and kept him under guard. Brown's last attempt to disrupt the American approach was to dig a mine from the fort to an abandoned house that he believed the Americans would use as a shelter for their final assault on the fort. Lee described the close brush with death a party of American riflemen had as a result of Brown's efforts.

> *About three in the morning of the fourth of June, we were aroused by a violent explosion, which was soon discovered to have shattered the very house intended to be occupied by the rifle party before daybreak. It was severed and thrown into the air thirty or forty feet high, its fragments falling all over the field.... Browne* [had] *pushed a sap* [tunnel] *to this house,*

---

[42] Lee, 362
[43] Ibid., 363

> which he presumed would be certainly possessed by the besieger, when ready to strike his last blow; and he concluded from the evident maturity of our works and from the noise made by the militia...that the approaching morning was fixed for the general assault. [44]

The American riflemen had come within an hour or two of obliteration.

Although Lieutenant Colonel Brown's actions caught the Americans by surprise, it did nothing to derail their plans for a final assault on Fort Cornwallis. Staring certain defeat in the face, Brown finally asked for surrender terms and relinquished the fort on June 5th.

Siege of Ninety Six

Lieutenant Colonel Lee and his Legion departed Augusta the day after Brown surrendered and joined General Greene outside of the last major British outpost in the South Carolina backcountry, Ninety Six. Greene had commenced a siege of Ninety Six in late May and progress was frustratingly slow. The outpost was a key British stronghold that had started out thirty years before the war as a trading post with the Cherokee Indians. Its name was a reference to the number of miles that colonists believed separated the outpost from Keowee, a prominent Cherokee Indian village.[45] Secured by the British in June 1780, Ninety Six became a vital post in a chain of forts used to control the South Carolina backcountry.

---

[44] Ibid., 366
[45] Robert M. Dunkerly and Eric K. Williams, *Old Ninety Six: A History and Guide*, (Charleston: SC: A History Press, 2006), 65

Ninety Six

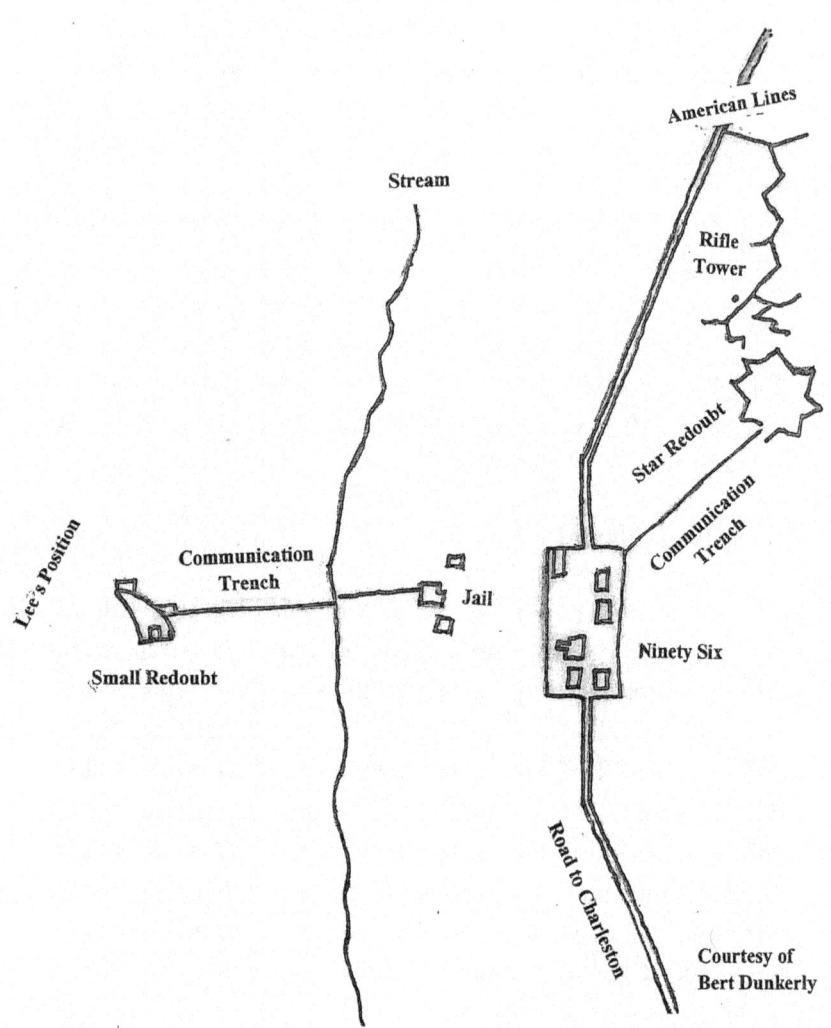

The fortifications around Ninety Six were significantly strengthened by the British prior to General Greene's arrival. A sturdy palisade wall of upright logs surrounded the rectangular shaped village of about a dozen buildings.[46] Many of the buildings served as fortified block houses and a deep ditch and thick layer of abattis encircled the village.[47] Each of these obstacles provided a layer of defense that attackers had to overcome. Just outside the western wall of the village stood a three story brick jail. The jail overlooked a ravine through which the water supply for the village flowed and was thus also strongly defended. A trench was dug from the jail down to the bottom of the ravine to allow the garrison to retrieve water under cover. On the opposite side of the ravine stood a small stockade fort that defended the western approach to the village and its water supply.[48] The defenses of this position were similar to the village and included a wooden palisade wall, block houses, an earthen parapet, deep ditch and abattis. A communication trench linked this position with the village.[49]

The strongest point of defense for Ninety Six stood over 100 yards outside the northeast corner of the village. A large, eight pointed star redoubt with 3 three pound cannon, steep fourteen foot high earthen walls, a deep moat and thick abattis guarded the northern and eastern approaches to the village.[50] The redoubt was connected to the village with a three and a half foot deep communication trench.[51]

---

[46] Ibid.
[47] Ibid.
[48] Ibid., 49-51
[49] Ibid.
[50] Ibid., 41
[51] Ibid., 41, 66

The entire post at Ninety Six was defended by 1,250 American Tories made up of approximately 400 experienced New York and New Jersey provincial troops who were equal to any British regulars and 850 Loyalist militia from Georgia and the Carolinas.[52] Lieutenant Colonel John Cruger, a prominent New York Loyalist and experienced provincial officer, commanded the outpost. Unaware that Lord Rawdon had ordered Ninety Six to be evacuated (Rawdon's order had been intercepted by the Americans) Cruger was determined to hold the fort at all costs.

General Greene was equally determined to capture the important outpost and ordered his troops (approximately 1,000 continentals from Virginia, Maryland and Delaware and a small number of militia) to lay siege to Ninety-Six in late May under the direction of Colonel Thaddeus Kosciuszko, a military engineer from Poland.[53] Greene's first attempt to establish a trench line just seventy yards from Cruger's star redoubt ended in failure when an enemy sortie routed the Americans at bayonet point. Greene moved his efforts further back and began his first trench and artillery battery at a more conventional and safe 400 yards.

Siege operations under the broiling South Carolina sun was grueling work for Greene's men. Two weeks into the siege Greene described to Congress the difficulties and hardships his men endured.

---

[52] Ibid., 66
[53] Ibid., 65

> *Our poor Fellows are worne out with fatigue, being constantly on duty every other Day and sometimes every Day. The [enemy] Works are strong and extensive. The position difficult to approach and the Ground extremely hard. The Garrison numerous and formidable when compared with our little force. They have sallied more or less every Night; but have been constantly driven in.*[54]

General Greene's frustration at his slow progress and mounting concern that Lord Rawdon might send a relief force to rescue Ninety Six prompted Greene to employ a few unorthodox siege methods. Borrowing a tactic used by Lieutenant Colonel Lee at Fort Motte, Greene shot flaming arrows (out of muskets) into Ninety Six in the hopes of burning the garrison out.[55] Lieutenant Colonel Cruger countered this by removing all of the roofs of the buildings.[56] Cruger also countered Greene's construction of a 30 foot tower / firing platform for riflemen just thirty yards from the star redoubt by raising the height of the redoubt's parapet with sand bags.[57]

Most of General Greene's efforts for the first two weeks of the siege centered on the star redoubt, but the arrival of Lieutenant Colonel Lee and his troops allowed Greene to begin pressuring the western approach to Ninety Six as well. While Lee's dragoons rode south under Lieutenant Colonel

---

[54] Conrad, ed., "General Greene to Congress, 9 June, 1781," *The Papers of General Nathanael Greene*, Vol. 8, 363
[55] Dunkerly and Williams, 37
[56] Stedman, 368-369
[57] Conrad, ed., "General Greene, to the President of Congress, 20 June, 1781," *The Papers of General Nathanael Greene*, Vol. 8, 419

Washington to help General Sumter and General Marion delay a 2,000 man British relief force from Charleston, Lee was ordered to, "*take post* [with his infantry] *opposite to the enemy's left, and to commence regular approaches against the stockade.*"[58] Once again Lee and his men found themselves engaged in siege warfare, and once again they, like their comrades advancing against the star redoubt, successfully fought off nightly attacks by the enemy and steadily advanced their trench line forward. General Greene described the relentless American efforts in another letter to Congress.

> *We had pushed on our approaches very near to the Enemys Works. Our third parallel [trench] was formed round their Abattis. A mine [tunnel] and two approach [trenches] were within a few feet of their Ditch. On the right [where Lee's Legion was] our approaches were very near the Enemys Redoubt.... We had raised several Batteries for Cannon, one upwards of 20 feet high within one hundred and forty yards of the Star Fort to command the Works, and a rifle Battery also within thirty yards to prevent the Enemy from annoying our Workmen. For the last ten Days not a Man could shew his head but he was immediately shot down, and the firing was almost incessant Day and Night.*[59]

The proximity of Lee's trench to the British stockade protecting Cruger's water supply and the intensity of Lee's small arms and artillery fire effectively reduced the garrison's

---

[58] Lee, 371-372
[59] Conrad, ed., "General Greene to Congress, 20 June, 1781," *The Papers of General Nathanael Greene*, Vol. 8, 419-421

access to water to a trickle by mid-June.[60] Despite the apparent success of the American efforts, however, time had run out for them. Neither General Sumter or Marion, or Lieutenant Colonel Washington were able to halt the advance of Lord Rawdon's relief force to Ninety Six. Rawdon had 2,000 British regulars with him and General Greene was reluctant to engage such a force with only half as many troops.[61] Reports on June 17th that Lord Rawdon was only thirty miles away convinced General Greene that he either had to end the siege or storm the fort in one last attempt to capture the garrison before Rawdon arrived. Greene chose the latter option and described the assault to Congress.

> *Lieut. Col. Lee and his Legion Infantry, and Captain* [Robert] *Kirkwoods light Infantry made the attack on the right, and Lieut Col.* [Richard] *Campbell with the first Maryland and the first Virginia Regiments was to have stormed the Star Redoubt, which is their principal Work, and stands upon the left. The parapet of this Work is near twelve feet high and raised with sand Bags near three feet more. Lieut.* [Isaac] *Duval of the Maryland Line, and Lieut* [Samuel] *Selden of the Virginia Line led on the forlorn hope, followed by a party with hooks to pull down the sand Bags....*
> *A furious Cannonade preluded the attack. On the right the Enemy was driven out of their Works and our people* [Lee and his men] *took possession of it.*

---

[60] Lee, 374-375
[61] Conrad, ed., "General Greene to Congress, 20 June, 1781," *The Papers of General Nathanael Greene*, Vol. 8, 419-421

> *On the left never was greater bravery exhibited than by the parties led on by Duval and Selden, but they were not so successful. They entered the Enemys Ditch and made every exertion to get down the sand Bags, which from the depth of the Ditch, height of the parapet, and under a galling fire, was rendered very difficult. Finding the Enemy defended their Works with great obstinacy and seeing but little prospect of succeeding without a heavy loss, and the issue doubtful, I ordered the attack to be pushed no further.*[62]

Lieutenant Colonel Lee and his force grudgingly abandoned the captured works at nightfall and returned to their own lines to prepare for the retreat from Ninety-Six.

Greene estimated that over 40 of his men were killed, wounded, or captured, mostly in the failed assault against the star redoubt. His losses for the entire siege numbered 185 men killed, wounded, and missing.[63] British losses were approximately half that amount.[64]

Convinced that his army was in significant danger from Lord Rawdon's force, Greene reluctantly lifted the siege and ordered a retreat northward towards Charlotte, North Carolina. After weeks of endless fatigue and struggle, Greene's troops were thoroughly demoralized by the turn of events. Lieutenant Colonel Lee recalled that, "*gloom and silence pervaded the American camp; every one disappointed – every*

---

[62] Ibid.
[63] Lee, 377
[64] Ibid.

one mortified. *Three days more, and Ninety-six must have fallen.*"[65]

Lord Rawdon reached Ninety Six on June 21st, two days after Greene's departure. Despite a shortage of provisions for his fatigued troops, Rawdon pressed on after Greene the next morning with a large portion of his force. Screening Greene's retreat was Lieutenant Colonel Lee and his legion along with Captain Robert Kirkwood's company of Delaware continentals. They encountered the vanguard of Rawdon's force at the Enoree River, but that was as far as Rawdon marched. Convinced that further pursuit was fruitless and a danger to his exhausted troops, Rawdon reversed direction and returned to Ninety Six to supervise the withdrawal of the garrison and loyalists.

Confident that the evacuees were adequately protected for their march to Charleston, Rawdon led approximately 850 men eastward towards Friday's Ferry on the Congaree River.[66] He planned to unite with a regiment of British regulars under Lieutenant Colonel Alexander Stewart, but unbeknownst to Rawdon, Stewart had turned back to Charleston.

When General Greene learned of these developments he redirected his march southward. He hoped to catch Rawdon before he was reinforced and instructed Lieutenant Colonel Lee and Lieutenant Colonel Washington to intercept and delay Stewart (whom Greene believed was still marching to the Congaree River). Unable to locate Stewart, Lee chose instead to harass Rawdon when the British commander withdrew from the Congaree River and marched southwestward to Orangeburg. Rawdon realized that the absence of Stewart

---

[65] Ibid., 378
[66] Stedman, 374

placed his detachment in a very vulnerable position so he marched towards Orangeburg in hopes of finding reinforcements. Although Lee was unable to halt Rawdon's march, a detachment of his cavalry under Captain Joseph Eggleston routed an enemy foraging party and captured 45 dragoons.[67]

Rawdon safely reached Orangeburg in early July and was soon joined by Lieutenant Colonel Stewart's wayward regiment. General Greene and his army also arrived in the vicinity of Orangeburg. Despite Rawdon's increased strength, estimated at 1,600 men to Greene's 2,000 men, Greene seriously considered an attack.[68] However, the strong defensive position of Rawdon and the hardship Greene's own men endured due to a lack of provisions, convinced General Greene to withdraw his army to the High Hills of the Santee, instead. According to Lieutenant Colonel Lee, the American army was in desperate need of rest and he approved of General Greene's decision. He recalled in his memoir that,

> *We had often experienced in the course of the campaign want of food, and had sometimes seriously suffered from the scantiness of our supplies, rendered more pinching by their quality; but never did we suffer so severely as during the few days' halt* [near Orangeburg]. *Rice furnished our substitute for bread, which, although tolerably relished by those familiarized to it from infancy, was very disagreeable to Marylanders and Virginians, who had grown up in the use of corn or wheat bread. Of meat we had*

---

[67] Lee, 381
[68] Ibid., 384

literally none; for the few meager cattle brought to camp as beef would not afford more than one or two ounces per man. Frogs abounded in some neighboring ponds, and on them chiefly did the light troops subsist. They became in great demand from their nutritiousness; and, after conquering the existing prejudice, were diligently sought after. Even the alligator was used by a few.... The heat of the season had become oppressive, and the troops began to experience its effects in sickness.[69]

These hardships along with the oppressive heat of the season, undoubtedly contributed to General Greene's decision to withdraw his army to the High Hills of the Santee to rest and recover.

---

[69] Ibid., 386-87

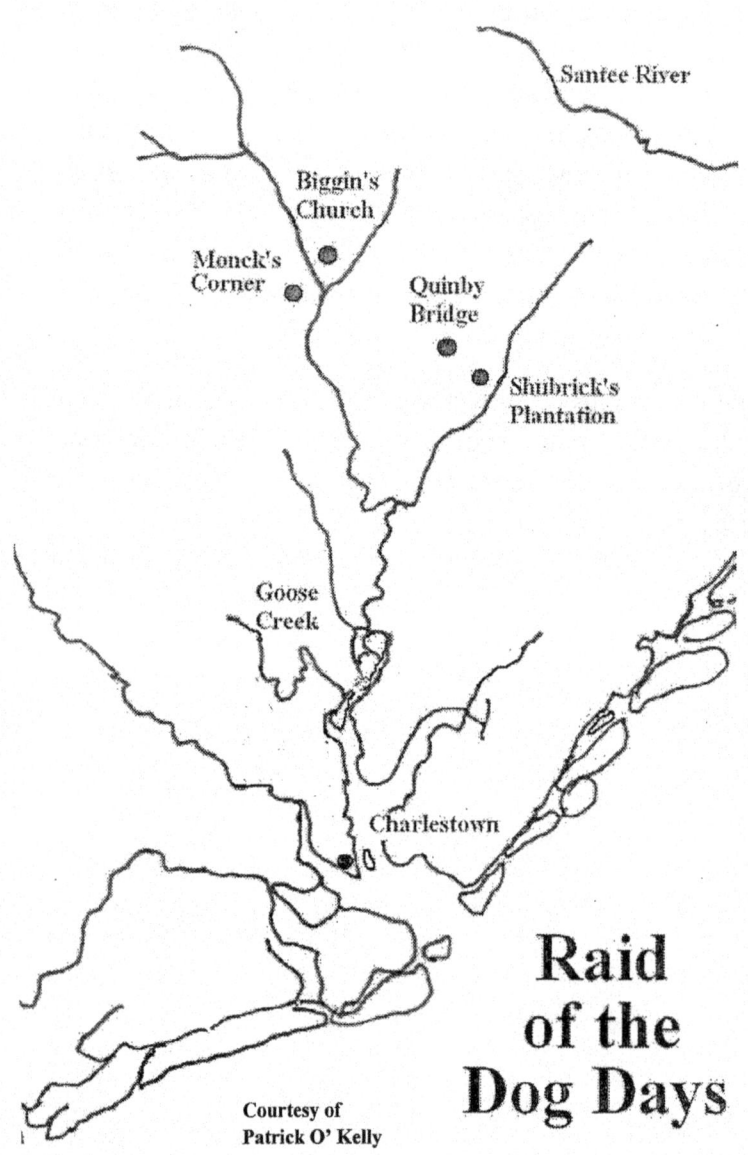

Chapter Ten

"Few Officers...are held in so high a point of estimation as you are."

While General Greene and the bulk of the American army withdrew to the High Hills of the Santee, Lieutenant Colonel Lee and his legion joined militia generals Thomas Sumter and Francis Marion in raids against enemy posts on the outskirts of Charleston. Greene hoped that such attacks so close to Charleston would force Rawdon to abandon Orangeburg and march south to protect the vital coastal town.[1] In doing so, more of South Carolina would be freed from British control.

After a few, *"trivial successes"* against British forces around Dorchester, Lee joined General Sumter and General Marion near Monks Corner and prepared to attack 600 British regulars and 150 mounted South Carolina Rangers under Lieutenant Colonel James Coates at Biggin's Church.[2] Recognizing the danger his outnumbered force faced, Coates avoided a fight and abandoned the post under cover of darkness, burning the church and whatever supplies he was unable to transport. Coats sent the mounted rangers to Charleston via one route and led his infantry and baggage train to the town via a different road.

---

[1] Conrad, ed., "General Greene to Congress, 17 July, 1781," *The Papers of General Nathanael Greene*, Vol. 9, 29

[2] Conrad, ed., "Headnote on the Dog Days Expedition," *The Papers of General Nathanael Greene*, Vol. 9, 13   and Lee, 387

The Americans ignored the fleeing rangers and pursued Coates and his force. Lee's dragoons, along with a detachment of South Carolina state cavalry under Lieutenant Colonel Wade Hampton, led the American pursuit. After an eighteen mile chase, Lee's cavalry fell upon Coates's rear guard and baggage train and captured it without a fight. Lieutenant Colonel Lee recalled that the inexperienced enemy rearguard, "*without discharging a single musket, threw down their arms and begged for quarters. Their supplication was cheerfully granted and like ourselves they escaped unhurt.*"[3]

The bulk of Lee's horsemen continued their pursuit of Coates and his main body. Captain John Armstrong commanded the lead section of Lee's dragoons and raced ahead to Quinby Bridge where he discovered the enemy posted on the other side. A party of British troops, covered by an artillery piece at the foot of the bridge, was loosening planks to prevent the Americans from crossing. Captain Armstrong halted and sent word back seeking instructions from Lee. Lieutenant Colonel Lee, not yet on the scene and unaware of the situation facing Armstrong, scolded the captain in his reply and reminded him that the orders of the day were to, "*fall upon the foe without respect to consequences.*"[4] Lee recalled that,

> *The brave Armstrong* [stung by Lee's answer] *put spur to his horse at the head of his section, and threw himself over the bridge upon the guard stationed there with a howitzer. So sudden was this charge that he drove all before him – the soldiers*

---

[3] Lee, 390
[4] Ibid.

> *abandoning their piece. Some of the loose planks were dashed off by Armstrong's section which, forming a chasm in the bridge, presented a dangerous obstacle. Nevertheless the second section, headed by Lieutenant Carrington, took the leap and closed with Armstrong, then engaged in a personal combat with Lieutenant Colonel Coates.... Most of* [the British] *soldiers, appalled at the sudden and daring* [American] *attack, had abandoned their colonel, and were running through the field, some with, some without arms, to take shelter in the* [nearby] *farm house.*[5]

The bold charge of Captain Armstrong and Lieutenant Carrington impressed more than just Lieutenant Colonel Lee. A British officer who witnessed it recounted:

> *We marched all that Night & a little after day break...crossed* [Quinby Bridge] *where we thought ourselves safe & took a little rest, but about nine OClock a party of Rebells galloped over the Bridge in the face of our Field Piece, rode through the Regiment, & wounded two Men: it was the most daring thing I ever heard of....*[6]

Lee and the rest of his dragoons soon arrived but were unable to leap the gap in the bridge (which had widened when Armstrong and Carrington's horsemen rumbled across). The inability of the rest of Lee's dragoons to cross the deep creek and join their comrades emboldened some of Lieutenant

---

[5] Ibid.
[6] Conrad, ed., "Headnote on the Dog Days Expedition," *The Papers of General Nathanael Greene*, Vol. 9, 15

Colonel Coates's soldiers to cease their flight and return to their steadfast commander.

Captain Armstrong and Lieutenant Carrington soon found themselves dangerously outnumbered and were forced to disengage and escape upstream. Once the British re-gained their artillery piece, Lieutenant Colonel Lee decided to withdraw from the bridge and wait for the infantry under General Sumter and General Marion to arrive.

Lieutenant Colonel Coates was determined to make a stand at a nearby farmhouse (known as Shubrick's plantation) and placed his men among the house and outbuildings for cover. Despite the enemy's strong position and his own lack of artillery, General Sumter, who assumed command of the American forces upon his arrival, ignored the counsel of General Marion and Lieutenant Colonel Lee to forsake a direct assault and renewed the attack. The American infantry, including Lee's Legion, forded the creek above the bridge and launched a furious assault on the plantation. General Marion commanded the left wing of the attack and recalled,

> *I marched my men to a fence about fifty yards from the Enimy under a very heavy fire, we soon made them take shelter in and behind the houses, but was fired on from the stoop of the Houses & through the doors, windows & Corners, our Ammunition being Intirely expended I was Obliged to retire.*[7]

General Marion's force suffered the most casualties in the failed American assault and he and his men were furious at Sumter's rash decision to attack. They left the expedition in

---

[7] Conrad, ed., "General Marion to General Greene, 19 July, 1781," *The Papers of General Nathanael Greene*, Vol. 9, 48

disgust that evening as did Lieutenant Colonel Lee and his legion.[8] With his force significantly depleted, General Sumter had no choice but to abandon his efforts and withdraw.

Lieutenant Colonel Lee and his tired legion marched northwest to join General Greene and the rest of the American army in their encampment in the High Hills of the Santee. The oppressive summer heat caused both sides to curtail operations for the rest of the summer. This allowed Greene's troops to restore their health and fighting effectiveness. Lieutenant Colonel Lee noted that,

> *The troops were placed in good quarters, and the heat of July rendered tolerable by the high ground, the fine air, and good water of the selected camp. Disease began to abate, our wounded to recover, and the army to rise in bodily strength.*[9]

The sultry days of mid-summer did not pass incident free for Lee and his legion, they continued to regularly patrol south of the High Hills. In one engagement, Lee and sixty of his dragoons boldly attacked an enemy convoy of 32 wagons escorted by 300 men.[10] Lee reported to General Greene that, "[We] *overwhelmed the cavalry escort and the van, but were forced to break off the attack when the main body of the* [escort] *remained regular and cool.*"[11]   General Greene

---

[8] Conrad, ed., "Headnote on the Dog Days Expedition," *The Papers of General Nathanael Greene*, Vol. 9, 16

[9] Lee, 393

[10] Conrad, ed., "Colonel Henry Lee, Jr., to General Nathanael Greene, 8 August, 1781," *The Papers of General Nathanael Greene*, Vol. 9, 150

[11] Ibid.

lightly chided Lee for his bold attack and reminded him how critical Lee and his legion were to the army.

> *The services of the Corps so multiplies my obligation, and increases the confidence of the Army, that I fear I shall never be able to discharge the obligations, or support the spirits of the Army should any misfortune attend you at a critical hour.*[12]

While General Greene's army recuperated in the High Hills of the Santee, Lord Rawdon followed the advice of his physicians and left the South to recover his health. Lieutenant Colonel Stewart assumed command of Rawdon's troops in the field while Lieutenant Colonel Nisbet Balfour served securely in Charleston as the overall British commander in South Carolina.

With the last days of summer approaching, General Greene prepared to resume the offensive. Although he had yet to win a major battle, the relentless efforts of General Greene and his army had wrestled control of the South Carolina backcountry from the British. Lieutenant Colonel Lee and his legion played a crucial role in accomplishing this and by the end of August Lee's troops, like the rest of Greene's army, were eager to resume their efforts against the British.

---

[12] Conrad, ed., "General Greene to Colonel Henry Lee, Jr., 9 August, 1781," *The Papers of General Nathanael Greene*, Vol. 9, 152

Battle of Eutaw Spring

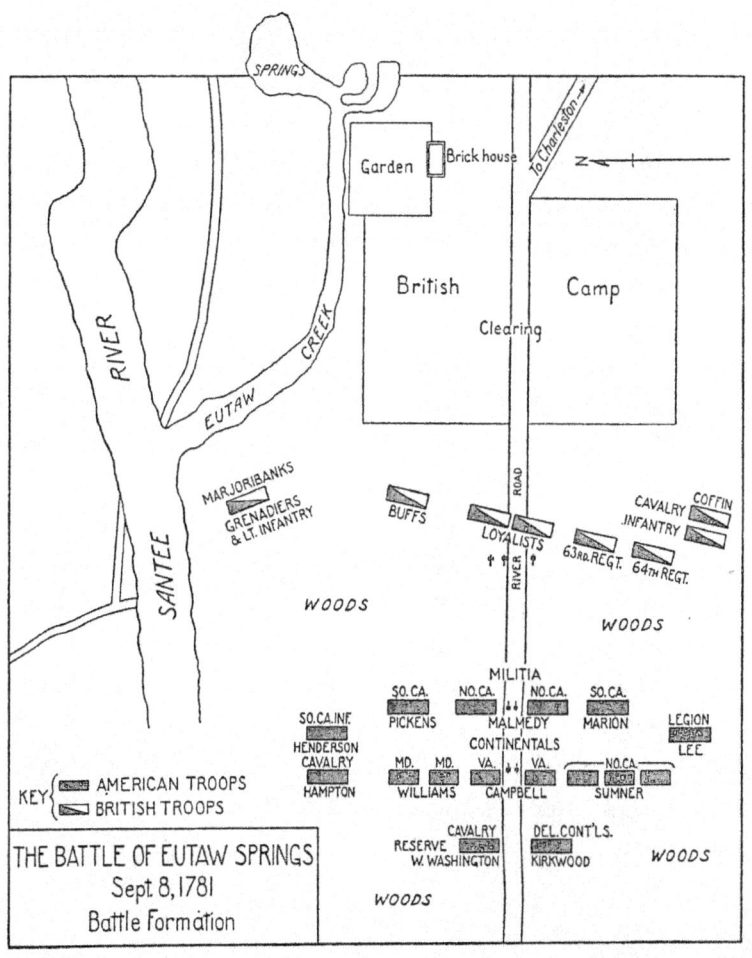

Source: Christopher Ward

Eutaw Springs

By early September, General Greene has assembled a force of over 2,000 continentals and militia troops at Motte's plantation on the Congaree River.[13] While there, Greene learned that a large British force over 2,000 strong under Lieutenant Colonel Stewart was encamped at Eutaw Springs, forty miles to the south.[14] Determined to prevent Stewart from establishing a permanent outpost at Eutaw Springs, Greene led his army south. They arrived unnoticed within a few miles of the enemy camp and prepared to attack early on the morning of September 8th.

General Greene's troops advanced towards the enemy at 4:00 a.m. with Lee's Legion and a detachment of South Carolina state troops under Lieutenant Colonel William Henderson in the lead. Lee and Henderson made contact with a party of enemy horse and infantry about four miles from Eutaw Springs. The 190 man Loyalist detachment under Major John Coffin apparently believed they had stumbled upon a party of American militia and made the mistake of charging the Americans.[15] Lieutenant Colonel Lee described the clash in a letter a few weeks after the battle.

---

[13] Conrad, ed., "Gen. Greene to the President of Congress, 11 September, 1781, Note 2," *The Papers of General Nathanael Greene*, Vol. 9, 333
[14] Conrad, ed., "Gen. Greene to the President of Congress 11 September, 1781," *The Papers of General Nathanael Greene*, Vol. 9, 328
[15] Conrad, ed., "Gen. Greene to the President of Congress, 11 September, 1781, Note 6-7," *The Papers of General Nathanael Greene*, Vol. 9, 334

> *We met a large body of the enemy foot with all their cavalry. They advanced on us & were met part way, we soon rendered it prudent for them to retire to their friends for support, which they performed in a great hurry with some loss.*[16]

General Greene confirmed Lee's account in a letter to Congress three days after the battle.

> [The enemy] *were soon convinced of their mistake by the reception they met with, the Infantry of the State Troops kept up a heavy fire, and the Legion in front under Captain Rudolph charged them with fixed Bayonets, they fled on all sides leaving four or five dead on the ground, and several more wounded.*[17]

At some point in this clash the Americans encountered a large number of British foragers in the area. The foragers, like Major Coffin's Loyalists, were routed and approximately 150 were captured.[18]

General Greene now decided to deploy his entire army for battle and formed his troops into two lines. The North and South Carolina militia troops under General Marion, General Pickens, and Colonel Francois Malmedy (a French volunteer officer) comprised General Greene's first line, while the North Carolina, Virginia, and Maryland continentals under General Jethro Sumner, Lieutenant Colonel Richard Campbell, and

---

[16] "Lieutenant Colonel Henry Lee Jr., to Unidentified, 2 October, 1781," Virginia Historical Society

[17] Conrad, ed., "Gen. Greene to the President of Congress, 11 September, 1781," *The Papers of General Nathanael Greene*, Vol. 9, 328

[18] Conrad, ed., "Gen. Greene to the President of Congress, 11 September, 1781, Note 7," *The Papers of General Nathanael Greene*, Vol. 9, 334

Colonel Otho Williams comprised Greene's second line. General Greene's four cannon were distributed between the two lines and Lieutenant Colonel Washington's cavalry and Captain Robert Kirkwood's Delaware continentals were held behind the second line in reserve. Lieutenant Colonel Lee's legion and Lieutenant Colonel Henderson's South Carolina state troops were ordered to cover the right and left flank of the army.[19] Both commanders shifted their men from the head of the column to the flanks of the militia line and advanced forward with the militia. They soon confronted Lieutenant Colonel Stewart's main body of approximately 2,000 troops deployed for battle across the road to Charleston.[20] Stewart's right flank was protected by a ravine and creek, but his left flank hung in the air and was vulnerable.

General Greene's attention, however, was focused on his militia line. He proudly described their conduct in battle in a letter to Congress.

> *All the Country is covered with Timber from the place the Action began to the Eutaw Springs. The firing began again between two and three miles from the British Camp. The Militia were ordered to keep advancing as they fired. The Enemies advanced parties were soon driven in, and a most tremendous fire began on both sides from right to left, and the Legion and State Troops were closely engaged. General Marion, Col. Malmady and General Pickens conducted the Troops with great gallantry and good*

---

[19] Conrad, ed., "Gen. Greene to the President of Congress, 11 September, 1781," *The Papers of General Nathanael Greene*, Vol. 9, 329

[20] Conrad, ed., "Gen. Greene to the President of Congress, 11 September, 1781, Note 2," *The Papers of General Nathanael Greene*, Vol. 9, 333

> conduct, and the Militia fought with a degree of spirit and firmness that reflects the highest honor upon this class of Soldiers.[21]

The militia's bold conduct was lauded by other American officers as well. Colonel Otho Williams noted,

> *It was with equal astonishment, that both the second line* [of continentals] *and the enemy,* [watched the American militia] *steadily, and without faltering, advance with shouts and exhortations into the hottest of the enemy's fire, unaffected by the continual fall of their comrades around them.*[22]

Even Lieutenant Colonel Lee acknowledged the militia's efforts. *"Our front line* [of militia] *behaved with great propriety & kept up a severe close fire...."*[23] Unfortunately, Lee observed that the militia was, *"Opposed with the utmost gallantry* [by the enemy] *& being liable to the bayonet, they got into disorder."*[24] With the militia line faltering, General Greene ordered the North Carolina continentals forward to support them. General Greene proudly noted that General Sumner's raw recruits,

---

[21] Conrad, ed., "Gen. Greene to the President of Congress, 11 September, 1781," *The Papers of General Nathanael Greene*, Vol. 9, 329

[22] Conrad, ed., " Gen. Greene to the President of Congress, 11 September, 1781, Note 8," *The Papers of General Nathanael Greene*, Vol. 9, 335

[23] "Lieutenant Colonel Henry Lee Jr., to Unidentified, 2 October, 1781," Virginia Historical Society

[24] Ibid.

> *Fought with a degree of obstinacy that would do honor to the best of veterans.... They kept up a heavy and well directed fire....*[25]

The rest of General Greene's continental troops also entered the fray at this point. Greene described the effect they had on the battle.

> *The Virginians under Lieut Col. Campbell, and the Maryland Troops under Col. Williams were led on to a brisk charge with trailed Arms, through a heavy cannonade, and a shower of Musquet Balls. Nothing could exceed the gallantry and firmness of both Officers and Soldiers upon this occasion. They preserved their order, and pressed on with such unshaken resolution that they bore down all before them. The Enemy were routed in all quarters. Lt. Col. Lee had with great address, gallantry, and good conduct, turned the Enemy's left flank, and was charging them in rear at the same time the Virginia and Maryland Troops were charging them in front.*[26]

Most of Lieutenant Colonel Stewart's men withdrew in disarray through their camp. Some took post inside a three story brick house, others behind garden walls and outbuildings. General Greene noted that his troops,

> *Kept close at the Enemy's heels after they broke until we got into their Camp, and a great number of*

---

[25] Conrad, ed. "General Greene to the President of Congress, 11 September 1781," *The Papers of General Nathanael Greene*, Vol. 9, 329
[26] Ibid., 331

> *Prisoners were continually falling into our hands, and some hundreds of the Fugitives run off towards Charles Town.*[27]

Not all of Stewart's men fled, however. A detachment of nearly 300 British infantry, under Major John Marjoribanks, maintained their post on the British right flank. Hidden among a dense thicket, the Americans initially overlooked and bypassed Marjoribanks and his men, but when Marjoribank's troops, "*wheeled around and* [attacked the Americans] *in the rear*," they sparked a good deal of consternation on the American left flank.[28] Lieutenant Colonel Washington led his cavalry in a bold charge to drive Marjoribanks back but discovered too late that the thicket that sheltered the British was virtually impenetrable. General Greene described what happened.

> *Lt. Col. Washington made the most astonishing efforts to get through the Thicket* [of Black Jack] *to charge the Enemy in the Rear, but found it impracticable, had his Horse shot under him, and was wounded and taken Prisoner.*[29]

Colonel Otho Williams added that,

> *A deadly and well directed fire…wounded or brought to the ground many of* [Washington's] *men and horses and every officer except two.*[30]

---

[27] Ibid.
[28] Stedman, 379
[29] Conrad, ed., "Gen. Greene to the President of Congress, 11 September, 1781," *The Papers of General Nathanael Greene*, Vol. 9, 331
[30] Conrad, ed., "Gen. Greene to the President of Congress, 11 September, 1781, Note 14," *The Papers of General Nathanael Greene*, Vol. 9, 336

Although the loss of Washington and his cavalry was a serious setback for the Americans, victory was still within their grasp. A spirited stand by Stewart's men in and around the brick house, however, took a heavy toll on Greene's men and stalled their advance. One British writer noted that,

> *Incessant peals of musquetry from the windows poured destruction upon the* [Americans], *and effectually stopped their further progress.*[31]

A violent struggle now erupted for possession of the brick house. Lieutenant Colonel Lee recalled that some of his own legion infantry nearly forced their way into the house.

> *The left of the Legion Infantry, led by Lieutenant Manning...followed close upon the enemy still entering* [the house], *hoping to force his way before the door could be barred. One of our soldiers actually got half way in, and for some minutes a struggle of strength took place – Manning pressing him in, and Sheridan* [the British commander in the house] *forcing him out. The latter prevailed, and the door was closed.*[32]

Trapped in the open outside of the house, Lieutenant Manning and his men reportedly used British prisoners to shield themselves as they withdrew.[33] Although Lee's men had failed to gain entrance into the house, General Greene

---

[31] Stedman, 379
[32] Lee, 470
[33] Conrad, ed., "Gen. Greene to the President of Congress, 11 September, 1781, Note 14," *The Papers of General Nathanael Greene*, Vol. 9, 336

remained determined to capture it and moved his cannon to within 100 yards of the structure.[34] This soon proved to be a mistake as British musket fire from the house and garden decimated the American gun crews. General Greene lauded the bravery of his doomed gun crews to Congress. *"Never were pieces better served, most of the Men and Officers were either killed or wounded."*[35]

Major Marjoribanks, who had withdrawn his men from the thicket to the garden walls near the house to better protect the British right flank, boldly charged the American cannon and seized them from the surviving artillerists. Marjoribanks was mortally wounded in the charge, but his sacrifice swung the battle in favor of the British. Colonel Ortho Williams claimed after the battle that the loss of the American cannon convinced General Greene to end the attack and withdraw.[36] General Greene attributed his decision to a number of factors.

> *Finding our Infantry galled by the fire of the Enemy, and our Ammunition mostly consumed, tho' both Officers and Men continued to exhibit uncommon acts of heroism, I thought proper to retire out of the fire of the House and draw up the Troops at a little distance in the Woods, not thinking it adviseable to push our advantage farther.... We collected all our Wounded, except such as were under the command of the fire of the House, and retired to the ground from*

---

[34] Conrad, ed., "Gen. Greene to the President of Congress, 11 September, 1781, Note 15," *The Papers of General Nathanael Greene*, Vol. 9, 337

[35] Conrad, ed., "Gen. Greene to the President of Congress, 11 September, 1781," *The Papers of General Nathanael Greene*, Vol. 9, 331

[36] Conrad, ed., "Gen. Greene to the President of Congress, 11 September, 1781, Note 16," *The Papers of General Nathanael Greene*, Vol. 9, 337

> which we marched in the morning, there being no Water nearer, and the Troops ready to faint with the heat, and want of refreshment, the Action having continued near four Hours.[37]

Despite their retreat and the heavy losses in men and officers (estimated at over 500 men for the Americans and 700 for the British) General Greene proclaimed the battle a victory for his army.[38] In his letters after the battle, Greene described the engagement as, *"by far the most bloody and obstinate I ever saw,"* and, *"a complete victory."*[39] And although he admitted that the fortunes of battle prevented his troops from taking, *"the whole british Army,"* Greene found vindication in the fact that Stewart withdrew to the vicinity of Charlestown soon after the battle.[40] Greene's army pursued but was unable to overtake Stewart before he reached Charlestown, so General Greene returned to the High Hills of the Santee to rest his tired army.

---

[37] Conrad, ed., "Gen. Greene to the President of Congress, 11 September, 1781," *The Papers of General Nathanael Greene*, Vol. 9, 332

[38] Tarleton, "Extract of a letter from Lieutenant-colonel Stewart to Earl Cornwallis, 9 September, 1781," 512 and
Conrad, ed., "Gen. Greene to the President of Congress, 11 September, 1781, Note 24," *The Papers of General Nathanael Greene*, Vol. 9, 338

[39] Conrad, ed., See letters from General Greene to:
    Governor Thomas Nelson, 16 September, 1781, 350
    Governor Thomas Burke, 17 September, 1781, 355
    Marquis de LaFayette, 17 September, 1781, 358
    General Washington, 17 September, 1781, 362
The Papers of General Nathanael Greene, Vol. 9

[40] Ibid.

Rest, Recruitment and Romance

While his troops rested, General Greene called upon the political leaders of the southern states, as well as Congress and even General Washington, to send more assistance to his army, but little was forthcoming. When one of Lee's officers returned to camp in late September from an unsuccessful recruiting mission in Virginia and reported that state officials were unable (Lee thought unwilling) to send new recruits and supplies southward, an exasperated Lee exclaimed,

> *What man of common fire, can submit to serve under such odious distinctions...when the* [demands] *of the service, & the* [harshness] *of the climate daily thin our ranks.*[41]

Lee worried that if the states continued to withhold new recruits for his legion his unit would very shortly be combat ineffective.

General Greene held similar concerns, but his were for the entire southern army. He wrote to General Washington on September 29th and declared, *"our situation is truly distressing, and the want of a reinforcement very pressing."* [42] Greene asserted that he did not want to undermine Washington's efforts at Yorktown, but if it were possible, he could really use reinforcements in South Carolina.[43] General Greene made a similar appeal to Washington a week later and

---

[41] Conrad, ed., "Colonel Henry Lee Jr., to General Greene, 28 September, 1781," *The Papers of General Nathanael Greene*, Vol. 9, 406-407
[42] Conrad, ed., "General Greene to General Washington, 29 September, 1781," *The Papers of General Nathanael Greene*, Vol. 9, 412
[43] Ibid.

sent Lieutenant Colonel Lee northward to deliver it and stress the need for help. In addition to delivering the letter, Lee was to confer with Washington about the commander-in-chief's plans for the southern theater of the war.[44]

Lee headed north on October 8th, and arrived in Yorktown a few days before Cornwallis surrendered on October 19th.[45] He briefed General Washington on the situation in South Carolina, witnessed the surrender ceremony at Yorktown, and spent a few days with family and friends.[46] One of his stops was at Stratford Hall in Westmoreland County where he paid a visit to his eighteen year old second cousin, Miss Matilda Ludwell Lee.

The "divine Matilda" as she was called by her admirers, was the eldest daughter of Philip Ludwell Lee and the heiress of Stratford Hall, a grand plantation overlooking the Potomac River.[47] The young couple had probably met as children years earlier during one of Harry Lee's visits to Westmoreland County with his father. It is also likely that Harry Lee visited Matilda in December 1780 on his way to join General Greene in South Carolina as he asked about her in a letter to his brother in April 1781.[48] The only other opportunity Lee had to visit Matilda during the war was in early 1780 while he was on a short leave from the army.

---

[44] Conrad, ed., "General Greene to the President of Congress, 7 October, 1781," *The Papers of General Nathanael Greene*, Vol. 9, 430
[45] Lee, 518
[46] Conrad, ed., "General Washington to General Greene, 31 October, 1781, *The Papers of General Nathanael Greene*, Vol. 9, 505
[47] Cazenove Gardner Lee Jr., *Lee Chronicle*, 85
[48] "Henry Lee to Charles Lee," 4 April, 1781," Lee Family Papers, 1638-1867, Virginia Historical Society

Although it is difficult to determine when the romance between Harry and Matilda started, it appears that upon Lee's return to the southern army in November 1781, he was determined to marry Matilda. Lee sought a leave of absence from General Greene just weeks after his return and by the spring of 1782 Harry Lee and Matilda were married.[49]

Lee Leaves the Army

Harry Lee's apparent engagement was not the only significant development in Lee's life in the weeks following Yorktown. Just three months after the allied victory, with the war seemingly drawing to a victorious close, Lieutenant Colonel Lee inexplicably left the army. Lee's rash action was apparently prompted by a deep disillusionment and perhaps depression that struck Lee in late 1781. The cause of Lee's sentiments is difficult to determine precisely. One possibility is that after six long years of hard service and heavy fighting, Lee was struck with battle fatigue or post traumatic stress. Another possibility might have been jealousy and anger at the appointment of Lieutenant Colonel John Laurens to Lee's Legion. Laurens, the son of a distinguished member of Congress and former aide to General Washington, joined Greene's army in December 1781 and was placed in charge of Lee's legion infantry.

At the time this occurred, General Greene was struggling to reorganize his army and many of his officers were upset by his decisions. No doubt many of Lee's officers, and perhaps Lee himself, seethed at the appointment of Laurens (an officer

---

[49] Conrad, ed., "General Greene to Colonel Henry Lee, Jr., 28 December, 1781," *The Papers of General Nathanael Greene*, Vol. 10, 126-127

with little combat experience) to the legion. To make matters worse, since Laurens held an older commission in the army, he was placed in command of the last significant military operation that Lee participated in, the aborted attack on Johns Island in mid-January 1782. Although Lee expressed his willingness to serve under Laurens prior to the attack, his request to leave the army came within days of the failed operation against Johns Island and a month after his first request for a leave of absence.

Anger and disappointment at General Greene almost certainly contributed to Lee's departure from the army as did attacks upon his character from fellow officers. Lee's letter to General Greene requesting permission to leave the army suggests that Lee was in great emotional distress, caused in part, by what he viewed as a lack of appreciation and recognition from General Greene and Lee's fellow officers.

> *I must at length ask permission to absent myself from the army. Disquietude of mind and infirmity of body unite in giving birth to my request. The first arises from the indifference with which my efforts to advance the cause of my country is considered by my friends, the persecution of my foes, and my consciousness that it is not in my power to efface the disagreeable impression. The second owes its birth to the fidelity with which I have served, and is nourished by my continuance in the same line of conduct. However disgusted I am with human nature, I wish, from motives of self, to make my way*

> *easy and comfortable. This, if ever attainable, is to be got only in an obscure retreat.*[50]

Lee's letter shocked and dismayed General Greene. He replied almost immediately.

> *I have beheld with extreme anxiety for some time past a growing discontent in your mind and have not been without my apprehensions that your complaints originated more in distress than in the ruins of your [health]. Whatever may be the source of your wounds I wish it was in my power to heal them.*[51]

General Greene also addressed Lee's accusation that certain officers, particularly Greene, had slighted his efforts and accomplishments and in doing so had aided Lee's detractors and harmed his public reputation.

> *You say your friends are not disposed to do justice to your exertions. If you mean me and anything appears in my conduct to confirm it, it has been owing to error in judgment or accident and not to a disinclination.... I am far from agreeing with you in opinion that the public will not do you justice or that they do not do you justice. I believe few Officers either in America or Europe are held in so high a point of estimation as you are.... It is true there are a few of your Country men who from ignorance and*

---

[50] Conrad, ed., "Colonel Henry Lee, Jr. to General Greene, 26 January, 1782," *The Papers of General Nathanael Greene*, Vol. 10, 264-265
[51] Conrad, ed., "General Greene to Colonel Henry Lee, Jr., 27 January, 1782," *The Papers of General Nathanael Greene*, Vol. 10, 268-269

> *malice are disposed to do injustice to your conduct but it is out of their power to injure you. Indeed you are ignorant of your own weight and influence otherwise you would despise their spleen and malice. There is no inconvenience I am not ready to submit to to oblige you nor is there any length I would not go to serve you in the line of truth and honor; but let me entreat you not to think of leaving the Army.*[52]

General Greene's letter had little effect on Lee; he was determined to leave the army. Three weeks passed during which Lee continued with his duties and worried about his men. He expressed the hope that his replacement was not, *"an experimenter* [else] *he will waste the troops very fast,"* and he worried about the condition of the legion's horses and the scarcity of forage for them.[53]

Lee also discussed the matter of his departure with a number of officers, including General Greene. No one could convince Lee to change his mind. A conversation Lee had with Colonel Lewis Morris was conveyed to General Greene and prompted a heated letter to Lee from Greene that shed light on at least part of what troubled Lee. Greene wrote,

> *I am exceeding sorry to find that notwithstanding all that has passed between us upon the subject of your discontent that your disgust increases and that you harbor sentiments respecting me no less groundless than unfriendly.... You say you think you have been injured by my public report of the battle of the*

---

[52] Ibid.
[53] Conrad, ed., "Colonel Henry Lee, Jr. to General Greene, 13 February, 1782," *The Papers of General Nathanael Greene*, Vol. 10, 360

> *Eutaws. Take up the matter collectively and separately and I am far from thinking so. In the course of the last campaign there were eleven public reports made, in all of which you are honorably mentioned. Even in my defeats you are spoke of with great respect. Where is there an Officer that stands in the least competition with you?*[54]

General Greene reminded Lee of the challenges he faced as the army's commander and how he always looked out for the interests of Lee and his legion because those interests were usually in unison with the interests of the army. Greene also reminded Lee about all of the slights and abuse he [Greene] had suffered over the course of the war and called Lee's complaints "groundless" in comparison. Greene then wrote something that must have really irritated Lee.

> *Col Laurens thinks you have no reason in the World to complain; and that you do injustice to your own importance to dwell upon single expressions. All I ask of you is, to give yourself time to cool, and to take a general view of the Southern operations and see how important a part you have played in them.*[55]

Greene continued with a strong declaration of his friendship for Lee and a denial that he had displayed any favoritism towards Lieutenant Colonel Laurens (General Washington's former aide) when he appointed Laurens to command Lee's infantry.

---

[54] Conrad, ed., "General Greene to Colonel Henry Lee, Jr., 18 February, 1782," *The Papers of General Nathanael Greene*, Vol. 10, 378-379
[55] Ibid.

> *I love and esteem you, and wish you not to think meanly of me, as some of your insinuations seem to impart, that to compliment General Washington I had done injustice to you* [by appointing Laurens]. *As nothing is more remote from truth, so nothing is more wounding to my feelings.... I esteem General Washington but I should abhor myself, if I was capable of so dishonorable a sacrifice.*[56]

General Greene concluded his letter by disputing the allegation that he had shown favoritism to Colonel William Washington in his report on the Battle of Eutaw Springs.

> *I have a regard for Col Washington and think him a good Officer. But I don't mean to draw a comparison between you and him. Everybody knows my sentiments on that head.*[57]

Lieutenant Colonel Lee replied to General Greene on the eve of his departure with an apology for the grief he had caused the general and a denial of some of Greene's claims. In doing so, Lee revealed that he did indeed believe that Greene was partially responsible for the, "*numerous aspersions*" leveled against Lee's character.

> *I am much mortified at the trouble which my stupid conduct gives you. I delivered myself personally & fully to you in conversation on my matters; I now repent what I conceive to be the sum of that conversation, viz, I felt myself much aggrieved at the*

---

[56] Ibid.
[57] Ibid.

> *very polite compliments which were paid in your public reports to every officer employed by you in the course of the campaign, but myself. I considered that this distinction gave ground for the numerous aspersions, with which my character stood loaded. At the same time, I knew it was to be imputed to [accident] & not to design, as I had such increasing experience of your partiality & love for me.*[58]

Lee went on to deny some of the insinuations General Greene referenced in his letter and assured Greene of his strong feelings of friendship for him. *"My friendship & high respect for you is as firm & warm this moment as it ever was at any period in my life."*[59] Lee also claimed that jealousy towards his fellow officers was not a factor in his feelings.

> *As to the characters of many officers in the Southern army, no man has a higher respect for some of them than myself. I never wished to hear their merits lessened, nor had I ever an ambition to be declared their superior. Therefore jealousy does not make up the causes of my grief. My ambition might have been very great once, it is not so now, & I rest contented with my crass future.*[60]

Lee concluded his letter pessimistically.

---

[58] Conrad, ed., "Colonel Henry Lee, Jr. to General Greene, 19 February, 1782," *The Papers of General Nathanael Greene*, Vol. 10, 389-391
[59] Ibid.
[60] Ibid.

> *I hope never more to say one word concerning my official character or private disquietudes – it is my misfortune; & it is doubly so as the man I most love in the world,* [General Greene] *has been the undesigned instrument in fixing misery on me. Time & philosophy may recover me, tho' if I may judge from my feelings for months past, I ought to dispond....I am so affected by my separating myself from those whom I esteem chief among mankind, & so distressed at your countencing reports so contrary to truth...& derogatory to my honor that I must conclude with begging your pardon for the incoherence of this letter.*[61]

And with those stinging comments, Henry Lee headed north, ending six years of distinguished military service. He had entered the conflict in 1776 as an inexperienced twenty year old youth of a prominent Virginia family. Within a year of joining the northern army in New Jersey, Lee's exploits as a daring cavalry commander caused the enemy to specifically target him for capture. Lee's successful resistance and escape at Scott's Farm in 1778 only added luster to his reputation, and shortly afterwards the 22 year old newly promoted major found himself in command of an independent partisan corps of dragoons.

Lee's daring exploits continued and included a bold night assault on Powles Hook in 1779 and a brave stand against the enemy at Springfield in 1780. By the time Lee was ordered south in late 1780, he had risen to the rank of Lieutenant

---

[61] Ibid.

Colonel and was in charge of a legion of 150 infantry and 150 dragoons.

It was in the southern theatre in 1781, fighting under his friend Nathanael Greene and alongside such militia heroes as Francis Marion and Andrew Pickens that Lee (and his legion) really shined. They played crucial roles in the Race to the Dan and Guilford Courthouse, and punished the enemy in numerous engagements including, Pyle's Hacking, Fort Watson, Fort Motte, Fort Granby, Fort Galphin, the siege of Augusta, (successful), the siege of Ninety Six (unsuccessful), Quinby Bridge, and Eutaw Springs. Again and again General Greene credited Lee and his legion for the essential service they played in driving the British back to Charleston.

The service and accomplishments of Lee and his Legion during this period were unparalleled in the army and one of the reasons the legion and its commander were universally known and revered in America. Few officers in the Revolutionary War served with greater distinction than "Light Horse" Harry Lee and there is little doubt that Lee made the right decision in 1778 when he declined General Washington's offer to join his staff with the declaration, *"I am wedded to my sword."*

Portrait of Light Horse Henry Lee 1782

Artist: Charles Wilson Peale

Epilogue

In his last letter to General Greene before he left the army in February 1782, Henry Lee III declared that he, *"hoped to never more...say one word concerning my official character or private disquietudes."*[1] Lee returned to Virginia and plunged into domesticity with his marriage to Matilda. Lee's friend and fellow officer, General George Weedon of Fredericksburg, informed General Greene in November 1782 that, "[Lee] *has Pitched his Tent at Stratford in the Northern Neck.*"[2]

Although Lee refused to discuss his departure with General Greene any further, his silence apparently did not extend to other officers in the army. In the fall of 1782, General Greene learned of Lee's continued anger about Greene's reports on Eutaw Springs. In yet another attempt to assure Lee that he intended no harm to his reputation and held Lee in the highest regard, General Greene wrote one last letter to him on the topic of Eutaw Springs.

> *I see by your letter to Major Burnet you think great injustice was done your Legion in the report of the battle of Eutaw; and you lament my giving credit to the idle stories in forming my report. You may rest assured I did not. There was no man that deserved*

---

[1] Conrad, ed., "Colonel Henry Lee, Jr. to General Greene, 19 February, 1782," *The Papers of General Nathanael Greene*, Vol. 10, 389-391

[2] Denis Conrad, ed., "General George Weedon to General Nathanael Greene, 6 November, 1782," The Papers of General Nathanael Greene, Vol. 12, (Chapel Hill: University of NC Press, 2002), 154-155

> *greater credit than you that day; and if you are not so represented, it is my fault.*[3]

General Greene explained that by highlighting the deeds of the Legion infantry at Eutaw Spring under Major Michael Rudolph, he did not mean to diminish the efforts of the Legion cavalry commanded by Lee.

> *The infantry of the Legion deserved everything that could be said of them also, nor was the Cavalry blameable but less fortunate. They did not make a successful charge in the course of the day tho they attempted it several times. Two Corps may be equally disposed to distinguish themselves one may have an opportunity the other not, and where a case of this kind should happen I would ask you whether you could report them to equal advantage? If they can be then intention is everything and action nothing. I meant to speak of things as they happened, but it was far from my heart to leave the least implication for censure against your Corps, much less against you.*[4]

---

[3] Conrad, ed., "General Greene to Henry Lee, Jr., 7 October, 1782," *The Papers of General Nathanael Greene*, Vol. 12, 40

Note: Colonel Samuel Hammond who commanded a detachment of South Carolina militia cavalry, claimed that the British could have been totally routed if Lee had properly led his cavalry during the attack. Hammond contended that Lee was distracted by an attempt to assume command of the Virginia continentals from Lt. Col. Richard Campbell who had been mortally wounded early in the attack.

See: Conrad, ed., *The Papers of General Nathanael Greene*, Vol. 9, 335-36

[4] Ibid.

Once again General Greene complimented Lee's bruised ego by declaring,

> *No man in the progress of the Campaign had equal Merit with you nor is their one so represented.... But it is my opinion if there is any part of the public who refuse to you the merit you deserve from the reports that have been made* [then] *all the reports in the World would not remove their prejudices. Names have greater influence with mankind* [than people realize] *and I am confident if your reputation has felt any violence, it has been owing to your being a Lee and not from anything that was done or said.*[5]

General Greene's letter of October 1782 appears to be the last that was written between the two men on the matter.

With the Treaty of Paris and end of the war in 1783, Nathanael Greene joined the civilian ranks. Sadly, Greene enjoyed the pleasures of civilian life for only a short period of time. His life ended tragically in Georgia in 1786 when he succumbed to what most historians conclude was heat stroke.

Henry Lee learned of Greene's death while serving in the Continental Congress in New York. He took the news hard. Despite their quarrel, Lee revered Greene and had named his first born child Nathanael Greene Lee in 1784. Sadly, Lee's son had died in infancy, the first of many difficult losses Lee experienced at Stratford Hall. A few years later, in 1790, his beloved wife Matilda passed away.

---

[5] Ibid.

Lee struggled on, finding refuge from his grief in political affairs. He had served in the Continental Congress, the state legislature, and Virginia's Constitutional Ratification Convention prior to the loss of his wife. Two years after Matilda's death Lee was elected governor of Virginia. He married Anne Hill Carter of Shirley plantation during his third term as governor and the couple had five children including their youngest son, Robert Edward Lee.

Henry Lee's public service continued after he relinquished the officer of governor in 1794. He was re-instated in the U.S. army with the rank of Major General in 1794 and led troops westward to crush the Whiskey Rebellion. A term in the U.S. House of Representatives followed, during which Lee eulogized George Washington upon his death in 1799 with the immortal words, *"First in War, First in Peace, First in the Hearts of his Countrymen."*

Lee and his fellow Federalist Party members were swept aside by Thomas Jefferson and the Democratic-Republicans in the election of 1800. Lee now channeled his energies into business affairs. Unfortunately, he was a complete failure and squandered most of his wealth on land speculation and bad investments.

Destitute of money and forced to move his family into a small house in Alexandria, Virginia, Lee spent nearly two years in jail (1809-11) fending off his creditors. He used his time productively, gathering material and writing his account of the southern campaign of the Revolutionary War. Lee hoped that the proceeds from the sales of his book would straighten out his financial situation, but it was not to be.

Released from jail in 1810, Lee eventually ended up in Baltimore on the eve of the War of 1812. He opposed the impending war with Britain and was severely beaten by a mob who found such views treasonous. Disillusioned by such treatment and desperate to recover his health, Lee sailed alone to the West Indies, where he spent five long years. In 1818 with his health deteriorating, Lee decided to return to America. Unfortunately, he became gravely ill on the voyage home and disembarked at Cumberland Island, Georgia. He spent his last days at the home of the daughter of his old commander, General Nathanael Greene.

Major General Henry Lee III passed away on March 25, 1818. He received a military funeral and was interred on Cumberland Island, where he rested until 1913 when his remains were moved to the chapel of Washington and Lee University in Lexington, Virginia to lie next to those of his youngest son, Robert E. Lee. It seemed only fitting that both men, so obviously wedded to their swords as they were, were united in their final resting place.

Bibliography

Abbot, W.W., et al eds. *The Papers of George Washington: Colonial Series, Vol. 1-10.* Charlottesville: University Press of Virginia, 1983-1995.

Babits, Lawrence E. *A Devil of a Whipping: The Battle of Cowpens.* Chapel Hill : The University of North Carolina Press, 1998.

Babits, Lawrence E. and Joshua B. Howard, *Long, Obstinate, and Bloody: The Battle of Guilford Courthouse*, Chapel Hill, NC: The University of North Carolina Press, 2009.

Baker, Thomas *Another Such Victory: The Story of the American Defeat at Guilford Courthouse That Helped Win the War for Independence.* Ft. Washington, PA: Eastern National, 2005.

Ballagh, James C. ed., *Letters of Richard Henry Lee,* Vol. 1-2. New York : Macmillan Co., 1911.

Boatner III, Mark M. *Encyclopedia of the American Revolution.* 3rd ed., Stanpole Books, 1994.

Bodle, Wayne. *The Valley Forge Winter: Civilians and Soldiers in War*. PA: Pennsylvania State University Press, 2002.

Boyd, Julian. ed. *The Papers of Thomas Jefferson,* Vol. 1-6. Princeton, NJ: Princeton University Press, 1950-52.

Boyle, Joseph Lee. *Writings from the Valley Forge Encampment of the Continental Army*. Vol. 1- 5 Bowie: Heritage Books Inc., 2000-2005.

Buchanan, John. *The Road to Guilford Courthouse: The American Revolution in the Carolinas*. NY: John Wiley & Sons, Inc., 1997.

Campbell, Charles, ed., *The Bland Papers: Being a Selection from the Manuscripts of Colonel Theodorick Bland Jr. of Prince George County, Virginia*. Petersburg: Edmund & Julian Ruffin, 1840.

Campbell, Charles. *The Orderly Book of that Portion of the American Army stationed at or near Williamsburg, Virginia under the command of General Andrew Lewis, from March 18th, 1776 to August 20th, 1776*. Richmond, VA: 1860.

Carrington, Henry B. *Battles of the American Revolution*. New York: A. S. Barnes & Co., 1877.

Cecere, Michael. *An Officer of Very Extraordinary Merit: Charles Porterfield and the American War for Independence, 1775-1780.* Westminster, MD: Heritage Books, 2004.

Cecere, Michael. *Captain Thomas Posey and the 7th Virginia Regiment.* Westminster, MD: Heritage Books, 2005.

Cecere, Michael. Great Things Are Expected from the Virginians: Virginia in the American Revolution. Westminster, MD: Heritage Books, 2008

Cecere, Michael. *In This Time of Extreme Danger: Northern Virginia in the American Revolution.* Westminster, MD: Heritage Books, 2006.

Cecere, Michael. *They Are Indeed a Very Useful Corps: American Riflemen in the Revolutionary War.* Westminster, MD: Heritage Books, 2006.

Cecere, Michael. *They Behaved Like Soldiers: Captain John Chilton and the Third Virginia Regiment.* Westminster, MD: Heritage Books, 2004.

Chase, Philander D. et al ed. *The Papers of George Washington: Revolutionary War Series.* Charlottesville: University Press of Virginia, 1985-2010.

Clark, William, et al eds. *Naval Documents of the American Revolution,* Vol. 1-11. Washington D.C.: 1964-2005.

Commager, Henry Steele. *Documents of American History*, New York: Appleton-Century-Crofts, 1963.

Commager, Henry and Richard Morris, ed. *The Spirit of 'Seventy-Six: The Story of the American Revolution as Told by Participants*. NY: Castle Books, 1967.

Cresswell, Nicholas. *The Journal of Nicholas Cresswell; 1774-1777*. New York: The Dial Press, 1924.

Cullen, Charles and Herbert Johnson, ed. *The Papers of John Marshall, Vol. 1*. Chapel Hill : Univ. of NC Press, 1974.

Dann, John C. *The Revolution Remembered: Eyewitness Accounts of the War Independence*. Chicago: University of Chicago Press, 1980.

Davies, K.G. ed. *Documents of the American Revolution: 1770-1783, Vol. 9*. Irish University Press, 1975.

Decher, Peter. ed., *Journal of Lt. William Feltman of the First Pennsylvania Regiment, 1781-1782*. Samen, NH: Ayer Co., 1969.

Dorman, John Frederick. *Virginia Revolutionary Pension Applications, Volumes 1-52*. Washington D.C., 1958-1995.

Draper, Lyman C. *King's Mountain and Its Heroes: History of the Battle of King's Mountain*. Cincinnati: Peter G. Thomson, 1881.

Dunkerly, Robert M. and Eric K. Williams, *Old Ninety Six: A History and Guide*, Charleston: SC: A History Press, 2006.

Dwyer, William M. *This Day is Ours! An Inside View of the Battles of Trenton and Princeton: November 1776-January 1777.* New Brunswick, NJ: Rutgers, University Press, 1983.

Ewald, Captain Johann. *Diary of the American War: A Hessian Journal.* New Haven: Yale Univ. Press, 1979. Translated & edited by Joseph Tustin.

Farish, Hunter Dickinson, ed., *Journal and Letters of Philip Vickers Fithian: A Plantation Tutor of the Old Dominion, 1773-1774,* Charlottesville: University of Virginia Press, 1999

Fischer, David Hackett. *Washington's Crossing.* Oxford University Press, 2004.

Fitzpatrick, John C. ed. *The Writings of George Washington from the Original Manuscripts, 1745-1799.* Vol. 15-25. Washington: U.S. Govt. Printing Office, 1931-.44

Force, Peter. ed., *American Archives: 5th Series.* Washington D.C.: U.S. Congress, 1848-1853.

Ford, Worthington C., et al eds., *Journals of the Continental Congress,* Vol. 1-27, U.S. Government Print Office, 1904-37

Graham, James. *The Life of General Daniel Morgan.* Bloomingburg, NY: Zebrowski Historical Services, 1993. Press of Virginia, 1965.

Graham, Major William, *General Joseph Graham and His Papers on North Carolina Revolutionary History*, Raleigh, NC: Edwards & Broughton, 1904.

Gwathmey, John, H., *Historical Register of Virginians in the Revolution*, Richmond, VA: Dietz Press, 1938.

Hamilton, Stanislaus M. ed. *Letters to Washington & Accompanying Papers, Vol. 5.* Boston & New York: Houghton, Mifflin, Co., 1902.

Hartmann, John W. *The American Partisan: Henry Lee and the Struggle for Independence: 1776-1780.* Burd Street Press, 2000

Heckert, C.W. *A German-American Diary: Notes of Related Historical Interest, Including Translated Excerpts from the Wiederholdt Diary.* Parsons, WV : McClain Printing Co., 1980.

Hening, William W. *The Statutes at Large Being a Collection of all the Laws of Virginia, Vol. 9.* Richmond: J. & G. Cochran, 1821.

Higginbotham, Don. *Daniel Morgan: Revolutionary Rifleman.* Chapel Hill: University of North Carolina Press, 1961.

Hume, Ivor Noel. *1775: Another Part of the Field.* New York: Alfred A. Knopf, 1966.

Idzerda, Stanley J., ed., *Lafayette in the Age of the American Revolution: Selected Letters and Papers*, Vol. 1-3, Cornell University Press, 1980.

Jackson, Donald, ed., *The Diaries of George Washington*, Vol. 2-3, Charlottesville: University Press of Virginia, 1976- 1978.

Jackson, John W. *Valley Forge: Pinnacle of Courage*. Gettysburg, PA: Thomas Publications, 1992.

Johnson, Henry P. *The Campaign of 1776 Around New York and Brooklyn*. New York: Da Capa Press, 1971.

Johnson, Henry, P., *The Storming of Stony Point on the Hudson, Midnight, July 15, 1779: Its Importance in the Light of Unpublished Documents*, New York: James T. White, 1900

Johnson, Henry P. *The Yorktown Campaign and the Surrender of Cornwallis:* 1781. Eastern National, 1997. Originally printed in 1881.

Kapp, Friedrich. *The Life of Frederick William von Steuben*. NY: Corner House Historical Publications, 1999. (Originally published in 1859)

Ketchum, Richard M. *Saratoga,: Turning Point of America's Revolutionary War*. NY: Holt & Co., 1997.

Ketchum, Richard M. *The Winter Soldiers.: The Battles For Trenton and Princeton*. New York: Henry Holt Co., 1973.

Lamb, Roger. *An Original and Authentic Journal of Occurrences During the Late American War from Its Commencement to 1783.* Dublin: Wilkinson & Courtney, 1809.

Lee Jr., Cazenove Gardner, *Lee Chronicle: Studies of the Early Generations of the Lees of Virginia,* New York: Vantage Press, 1957

Lee, Charles. *The Lee Papers, Vol. 1.* Collections of the New York Historical Society, 1871.

Lee, Edmund Jennings, *Lee of Virginia, 1642-1892: Biographical and Genealogical Sketches of the Descendants of Colonel Richard Lee,* Baltimore: Genealogical Publishing Co., 1983
Originally published in 1895

Lee, Henry. *The Revolutionary War Memoirs of General Henry Lee.* New York: Da Capo Press, 1998.
Originally Published in 1812.

Lesser, Charles H. ed. *The Sinews of Independence: Monthly Strength Reports of the Continental Army.* Chicago: The University of Chicago Press, 1976.

Loprieno, Don, *The Enterprise in Contemplation: The Midnight Assault of Stony Point,* Westminster MD: Heritage Books, 2004.

Lumpkin, Henry. *From Savannah to Yorktown: The American Revolution in the South.* New York: toExcel, 1987.

Marshall, John. *The Life of George Washington, Vol. 1-2.* Fredericksburg, VA: The Citizens Guild of Washington's Boyhood Home, 1926.

Martin, Joseph Plum. *Private Yankee Doodle.* Eastern Acorn Press, 1962.

Mays, David John, ed., *The Letters and Papers of Edmund Pendleton, Vol. 1.* Charlottesville: University Press of Virginia, 1967.

McGuire, Thomas, J., *The Philadelphia Campaign,* Vol. 1 Stackpole Books, 2006

McGuire, Thomas. *The Surprise of Germantown: October 4, 1777.* Cliveden of the National Trust for Historic Preservation, 1994.

McILwaine, H. R. ed., *Journals of the Council of the State of Virginia, Vol. 1.* Richmond, 1931.

Moore, Frank. *Diary of the American Revolution, from Newspapers and Original Documents.* 2 vols. New York: Charles Schibner, 1860. Reprint. New York: New York Times & Arno Press, 1969.

O'Kelly, Patrick. *Nothing But Blood and Slaughter: The Revolutionary War in the Carolinas.* Vol. 2-3 Blue House Tavern Press, 2004-2005.

Palmer, William. *Calendar of Virginia State Papers, Vol. 1- 2.* Richmond: James E. Goode, 1881.

Rankin, Hugh F. *The War of the Revolution in Virginia.* Williamsburg, VA: Virginia Independence Bicentennial Commission, 1979.

Reed, John F. *Campaign to Valley Forge: July 1, 1777 – December 19, 1777.* Pioneer Press, 1980.

Reed, William B. *Life and Correspondence of Joseph Reed,* Vol. 2, Lindsay and Blakiston: Philadelphia, 1847

Richardson, William H., *Washington and the Enterprise Against Powles Hook,* Jersey City, NJ: New Jersey Title Guarantee and Trust Co., 1929.

Rowland, Kate Mason. *The Life and Correspondence of George Mason, Vol. 1.* New York: Russell & Russell, 1964.

Rutland, Robert A. ed., *The Papers of George Mason, Vol. 1.* University of North Carolina Press, 1970.

Ryan, Dennis P. *A Salute to Courage: The American Revolution as Seen Through Wartime Writings of Officers of the Continental Army and Navy.* NY: Columbia University Press, 1979.

Saffell, W.T.R. *Records of the Revolutionary War, 3rd ed.* Baltimore: Charles Saffell, 1894.

Sanchez-Saavedra, E.M. *A Guide to Virginia Military Organizations in the American Revolution, 1774-1787.* Westminster, MD: Willow Bend Books, 1978.

Scheer, George F., and Hugh F. Rankin. *Rebels & Redcoats: The American Revolution through the Eyes of Those Who Fought and Lived It.* New York: Da Capo Press, 1987.

Scribner, Robert L. and Tarter, Brent (comps). *Revolutionary Virginia: The Road to Independence,* Volumes 1-7. Charlottesville: University Press of Virginia, 1978-1983.

Selby, John. *The Revolution in Virginia: 1775-1783.* Williamsburg, VA: The Colonial Williamsburg Foundation, 1988.

Sellers, John R. *The Virginia Continental Line.* Williamsburg: The Virginia Bicentennial Commission, 1978.

Showman, Richard K. *The Papers of General Nathanael Greene. Vol. 7-11.* Chapel Hill: University of North Carolina Press, 1997-2000.

Simcoe, Lt. Col. John. *Simcoe's Military Journal: A History of the Operations of a Partisan Corps Called the Queen's Rangers, Commanded by Lieut. Col. J. G. Simcoe, During the War of Revolution.* New York: New York Times and Arno Press, 1968.

Smith, Paul H. ed., *Letters of Delegates to Congress: 1774-1789.* Washington, D.C.: Library of Congress, 1976.

Smith, Samuel. *The Battle of Brandywine.* Monmouth Beach, NJ: Philip Freneau Press, 1976.

Smith, Samuel. *The Battle of Princeton.* Monmouth Beach, NJ: Philip Freneau Press, 1967.

Sparacio, Ruth and Sam, ed., *Virginia County Court Records: Deed and Will Abstracts of Westmoreland County, Virginia: 1747-48*, McLean, VA: The Antient Press, 1996.

Sparks, Jared. ed. *The Correspondence of the American Revolution being Letters of Eminent Men to George Washington, Vol. 2.* Boston : Little, Brown & Co., 1853.

Stedman, C. *The History of the Origin, Progress, and Termination of the American War, Volume 1 & 2.* London, 1794.

Stille, Charles. *Major-General Anthony Wayne and the Pennsylvania Line in the Continental Army.* Port Washington, NY: Kenniket Press, Inc., 1968. First published in 1893.

Stryatt, Harold. ed., "Alexander Hamilton to John Laurens, 11 September,1779," *The Papers of Alexander Hamilton*, Vol. 2, New York: Columbia University Press, 1962

Stryker, William. *The Battles of Trenton and Princeton.* Republished by The Old Barracks Association, Trenton NJ: 2001. (Originally published in 1898).

Stryker, William. *The Battle of Monmouth.* Princeton: Princeton University Press, 1927.

Symonds, Craig L. *A Battlefield ATLAS of the American Revolution*. The Nautical & Aviation Publishing Co. of America Inc., 1986.

Tarleton, Banastre. *A History of the Campaigns of 1780 and 1781 in the Southern Provinces of North America*. North Stratford, NH: Ayer Co., Reprinted, 1999 Originally printed in 1787.

Thacher, James. *A Military Journal during the American Revolutionary War*. Hartford: CT, S. Andrus and Son, 1854. Reprint, New York: Arno Press, 1969.

Townsend, Joseph. "Some Account of the British Army under the Command of General Howe, and of the Battle of Brandywine," *Eyewitness Accounts of the American Revolution*. New York: Arno Press, 1969.

Uhlendorf, Bernhard A. ed. & trans. *The Siege of Charleston: With an Account of the Province of South Carolina: Diaries and Letters of Hessian Officers*. Ann Arbor, MI: University of Michigan Press, 1938.

Wilkinson, James. *Memoirs of My Own Times, Vol. 1* Philadelphia: Abraham Small, 1816. Reprinted by AMS Press Inc., : NY, 1973

Willard, Margaret. ed., *Letters of the American Revolution: 1774-1776*. Boston & New York: Houghton Mifflin Co., 1925.

Wirt, William. *The Life of Patrick Henry*. New York: M'Elrath & Bangs, 1832.

Wright, Robert K. *The Continental Army.* Washington, D.C. Center of Military History: United States Army, 1989.

------------ *Journals of the Continental Congress.* Library of Congress Online at www.loc.gov

Periodicals

Boyle, Joseph Lee. "From Saratoga to Valley Forge: The Diary of Lt. Samuel Armstrong," *The Pennsylvania Magazine of History and Biography, Vol. 121, No. 3 July 1997.*

Dawson, Henry B. "General Daniel Morgan: An Autobiography," *The Historical Magazine and Notes and Queries Concerning the Antiquities, History and Biography of America. 2nd Series, Vol. 9.* Morrisania, NY, 1871.

Heth, William. "Orderly Book of Major William Heth of the Third (sic) Virginia regiment, May 15 – July 1, 1777," *Virginia Historical Society Collections, New Series, 11,* 1892.

Katcher, Philip. "They Behaved Like Soldiers: The Third Virginia Regiment at Harlem Heights", *Virginia Cavalcade,* Vol. 26, No. 2, Autumn 1976.

McMichael, James. "The Diary of Lt. James McMichael of the Pennsylvania Line, 1776-1778," *The Pennsylvania Magazine of History and Biography*. Vol. 16, no. 2. 1892.

Montresor, John. "Journal of Captain John Montresor," *The Pennsylvania Magazine of History and Biography. Vol. 5*. Philadelphia: The Historical Society of Pennsylvania, 1881.

Rees, John. *"What is this you have been about to day?" : The New Jersey Brigade at the Battle of Monmouth.* (2003).

Sergeant R. "The Battle of Princeton," *The Pennsylvania Magazine of History and Biography, Vol. 20, No. 1.* 1896.

Seymour, William. "Journal of the Southern Expedition, 1780-1783", *The Pennsylvania Magazine of History And* Biography, Vol. 7. 1883.

Sullivan, Thomas. "Before and After the Battle of Brandywine: Extracts from the Journal of Sergeant Thomas Sullivan of H.M. Forty-Ninth Regiment of Foot," *The Pennsylvania Magazine of History and Biography*. Vol. 31, Philadelphia: Historical Society of Pennsylvania, 1907.

Tyler, Lyon. "The Old Virginia Line in the Middle States During the American Revolution," *Tyler's Quarterly Historical and Genealogical Magazine: Vol.12*. Richmond, VA: Richmond Press Inc., 1931.

"Personal Recollections of Captain Enoch Anderson, an Officer of the Delaware Regiment in the Revoutionary War," *Papers of the Historical Society of Delaware, Vol. 16*. Wilmington: The Historical Society of Delaware, 1896.

-------- "Henry Lee to Charles Lee, 7 April, 1776," *Collections of the New York Historical Society*, Vol. 4, 1871,

Virginia Gazette.

    Purdie
    Dixon and Hunter
    Pickney
    Rind
    Dixon and NicolsonNew Jersey Gazette

Unpublished Works

"General Anthony Wayne to Major Henry Lee, Jr., 24 August, 1779," Historical Society of Pennsylvania

Goodwin, Mary. *Clothing and Accoutrements of the Officers and Soldiers of the Virginia Forces: 1775-1780*. 1962

Lee Family Papers, 1638-1867, Virginia Historical Society

    "Henry Lee to Charles Lee," 4 April, 1781,"

Lee-Ludwell Papers, Virginia Historical Society

"Henry Lee II to William Lee, 1 October, 1774,"

"Lieutenant Colonel Henry Lee Jr., to Unidentified, 2 October, 1781,"

McLane Papers, Vol. 1, Memoirs, New York Historical Society

Thomas Gates Papers, William L. Clements Library, Ann Arbor, MI

"Henry Lee III to Charles Lee, 5 July, 1775,"

Weedon, George. *Correspondence Account of the Battle of Brandywine, 11 September, 1777.* The original manuscript is in the collections of the Chicago Historical Society, Transcribed by Bob McDonald, 2001.

Index

1st Continental Light Dragoons, 34, 57, 69,
1st Maryland Regiment, 204-205, 215, 233
1st Virginia Convention, 15
1st Virginia Regiment, 233
2nd Maryland Regiment, 204, 214-215
2nd Virginia Convention, 18
2nd Virginia Regiment, 215
2nd Continental Light Dragoons, 37
3rd Continental Light Dragoons, 36, 89
3rd Virginia Convention, 22-23
3rd Virginia Regiment, 50-51
4th Continental Light Dragoons, 36
5th Virginia Convention, 24-26
11th Virginia Regiment, 76

Alexander, General William (Lord Stirling) 30, 47, 50, 108-109, 113, 115-116,
Alexandria, VA, 270
Angell, Col. Israel, 138
Armstrong, Gen. John, 47
Armstrong, Capt. Patrick, 165-166, 240-242
Arnold, Benedict, 140-142
Augusta, GA, 221-222
    siege of, 223-227
Balfour, Col. Nisbet, 244
Banister River, VA, 177
Baskenridge, NJ, 30
Baylor, Col. George, 36, 89-90, 92, 120
Beatty, Capt. William, 215
Belfried, John, 25-26, 35
Biggin's Church, SC, 239
Birmingham Meeting House, PA, 50-51
Bland, Theodorick
    Capt., 25
    Maj., 31, 33
    Col., 34-38, 42-44, 47, 50, 52, 57, 68, 92

Boston, MA, 14-16, 22
Boston Tea Party, 13-14
Bound Brook, NJ, 34, 38
Boyd's Ferry, VA, 164-165
Braddock, General William, 2
Brandywine, Battle of, 47-52
Brown, Col. Thomas, 223-227
Bunker Hill, MA, 21
Butler, Col. Richard, 66, 90, 101
Burgoyne, General John, 38, 61
Burlington, NJ, 92-93, 126
Burnett, Maj., 267
Burr, Aaron, 12
Caldwell, Hannah, 135
Camden, SC, 208, 210, 212-213, 217-219
  battle of, 144, 151, 196
Campbell, Lt. Col. George, 147-148
Campbell, Lt. Col. Richard, 214, 216, 233 247, 250
Campbell, Col. William, 185, 188, 191, 193, 199, 201

Canada, 140
Carnes, Capt. Patrick, 148, 150
Carpenter's Island, PA, 59
Carrington, Lt., 241-242
Carter, Anne Hill, 270
Carter, Robert, 13-14
Caswell, Gen. 189
Catawba River, NC, 158
Cecil Court House, MD, 46
Chads Ford, PA, 46-48, 51-52
Champe, John
  plot to capture Benedict Arnold, 140-142
Charlestown, MD, 81
Charleston, SC, 126, 144, 148, 151, 210, 219, 221, 232, 235, 239, 244, 248, 251, 254
  surrenders, 134
Charlotte, NC, 234
Chatham, NJ, 35
Chelton, Captain, 14
Cheroke Indians, 221, 227
Chester, PA, 52, 59, 66
Clark, Major John, 66, 111, 116-119

Clarke, Col. Elijah, 223-225
Clinton, George, 74
Clinton, Gen. Henry, 82-84, 86, 90-91, 98, 134, 136-137, 139-141
Coates, Col. James, 239-242
Coercive Acts, 14
Coffin, Maj. John, 246-247
College of NJ, 10, 12-13
Congaree River, SC, 219, 235, 246
Connecticut Farms, NJ, 135
Continental Congress, 4, 15-16, 25, 39, 43, 55, 73, 79-81, 98, 120, 123, 127, 132-133, 142, 144, 147, 151, 191, 212, 214, 230, 232-233, 247, 253, 257, 269-270
Cornwallis, Gen. Charles, 151-152, 158-159, 161-162, 164-165, 167-170, 172-173, 175-178, 183-184, 186-191, 193, 197-199, 201-203, 206-208, 213, 222

Cooches Bridge, DE, battle of, 44-46
Cowans Ford, NC, 159
Cowpens, SC, 150, 162
    battle of, 151-157
Craig, Captain, 60
Creek Indians, 222
Crewe, Major, 68
Cross Creek, NC, 189, 207
Cruger, Col. John, 230-232
Cumberland Island, GA, 271
D'Estaing, Admiral, 123-124
Dan River, VA, 175
    retreat to, 162-173
Danbury, CT, 88
Danville, VA, 165
Darby, PA, 62, 66
Davidson, Gen. William, 159
Deep River, NC, 183
Dinwiddie, Robert, 2
Dix's Ferry, VA, 164, 167
Dorchester, SC, 239
Dumfries, VA, 1
Duval, Lt. Isaac, 233-234

Eggleston, Capt. Joseph, 180, 236
Elizabethtown, NJ, 135, 137, 139
Englishtown, NJ, 124
Enoree River, SC, 235
Eutaw Spring, SC, 262, 267
  battle of, 246-255
Ewald, Captain Johann, 33, 45, 87
Fairfax County, VA, 16-17, 23
Fauquier, Francis 2
Febiger, Col. Christian, 101
Finley, Capt., 221
Fishkill, NY, 88
Fithian, Philip, 13
Ford, Lt. Col. 214-215
Forsyth, Capt. Robert, 81, 111
Fort Cornwallis, GA, 223-227
Fort Galphin, GA, 222
Fort Granby, SC, 219-221
Fort Grierson, GA, 223-225
Fort Mercer, PA, 58, 60
Fort Mifflin, PA, 58, 60

Fort Motte, SC, 219-221, 231
Fort Watson, SC, 225
  siege of, 210-212
Francisco, Peter, 205
Fredericksburg, NY, 88
Fredericksburg, VA, 81, 267
French & Indian War, 2, 9, 20, 22
Friday's Ferry, SC, 235
Gage, General Thomas, 15-16
Gates, Gen. Horatio, 144
Georgetown, SC, 210, 218
  attack on, 148, 150, 157
Gerard, Conrad Alexander, 95
Germain, Lord, 159
Germantown, PA, 57, 59
  battle of, 54-56
Gillies, James, 165-166
Gist, Col. 118
Graham, Joseph, Capt., 90 180, 185-186
Greene, General Nathanael, 42, 47, 51-52, 114, 117-118, 127-130, 137-139, 142, 144-145, 147, 150, 157-159, 161-164, 167,

169-170, 172, 175-178,
183-184, 186, 188-189,
191-192, 196-197, 199,
201, 203-205, 207-208,
210, 212-218, 221-222,
224, 227-228, 230-237,
239, 243-244, 246-264,
267-269, 271
Grey, Gen. Charles, 89
Guilford Court House, NC,
161-164, 189
battle of, 191-205
Gunby, Col. John, 204-205, 214-216
Hackensack, NJ, 109
Haddenfield, NJ, 61
Halifax Courthouse, VA, 177
Hamilton, Alexander, 53, 76, 115
Hampton, Col. Wade, 240
Hancock, John, 43
Hardy, Capt. Levin, 110
Harrison, Col. 214
Haw River, NC, 183-184, 189
Haws, Lt. Col. Samuel, 214-217
Head of Elk, MD, 42, 134
Heard, Lt. James, 192
Henderson, Lt. Col.
William, 246, 248
Henry, Patrick, 15, 18-20, 29, 81
High Hills of the Santee, SC, 217, 236-237, 239, 243-244, 255
Hillsborough, NC, 176-177, 179, 183
Hobkirk Hill, SC
battle of, 213-217
Horndon, Lt. William, 103
Houston, Samuel, 200
Howard, Lt. Col. John, 153, 156-157, 163, 205
Howe, General William, 29, 38-40, 43-48, 50-52, 54, 57-58, 62, 68-69,
Huger, Gen. Isaac, 161, 197
Hull, Lt. Col., William, 103
Intolerale Acts, 15
Irwin's Ferry, VA, 164-165, 168-169
Jameson, Maj. John, 69
Jefferson, Thomas, 175, 177, 187, 203, 270
Johns Island, SC,
attack, 258
Johnson, Thomas, 81
Kennett Square, PA, 46

Keowee, SC, 227
Kings Bridge, 90
Kirkwood, Capt. Robert, 233, 235, 248
Knyphausen, Gen. Wilhelm, 46, 48, 51-52, 91, 134-139
Kosciuszko, Col. Thaddeus, 230
LaFayette, Marquis de, 42-43, 61, 143
Lamb, Roger, 198
Laurens, Lt. Col. John, 115, 257-258, 261-262
Lawson, Gen. Robert, 196, 202
Lee, Charles, of Leeslyvania, 12
Lee, Gen. Charles, 20-24, 83
Lee, Francis Lightfoot, 1
Lee II, Henry, 1-4, 9-10, 13-18, 20

Lee III, Henry (Light Horse Harry),
  childhood, 9-10, 12-14
  Prince William County Independent Militia, 17
  aide to Gen. Charles Lee, 21-22, 24
  militia service, 23-24
  Virginia light dragoons, 24
  capt. of 5th Troop, 25-28
  joins Washington's army, 29-30
  at Bound Brook, 34
  uniforms for his troop, 35-36
  court martial, 40
  captures prisoners, 43
  role at Cooches Bridge, 46,
  observations on Brandywine, 50
  role at Brandywine, 52
  skirmish at Valley Forge, 53-54
  role at Germantown, 55
  patrols outskirts of Philadelphia, 57-59
  commanded by Gen. Washington, 60
  patrols NJ and PA, 61-62
  at Valley Forge, 65-71
  offered promotion to

General Washington's
staff, 76-77, 79
promoted to major, 80
partisan corps formed,
80-82
attached to Scott's light
infantry corps,
84-91
winter, 1778-79,
92-94
trains partisan corps,
95-97
annexes Capt. Allen
McLane's troop,
96-98
scouts Stony Point,
98-99
punishes deserters,
99-100
at Stony Point, 101
prepares attack on
Powles Hook, 105,
108
assault on Powles
Hook, 109-114
praised and criticized,
115-116
court martial, 116-
120
acquitted, 119-120
awarded medal from
Congress, 120-121
in New Jersey, 123-124,
126
forage detail, 131-132
furlough, 132
partisan Legion, 133
ordered south, 133
battle of Springfield,
135-139
plot to capture Benedict
Arnold, 141-142
partisan corps nearly
disbanded, 142
legion formed, 144
legion ordered south,
145
arrives in SC, 147, 151
unites with Gen. Marion,
148
attacks Georgetown, SC,
148, 150
ordered to rejoin army,
161
Race to the Dan, 163-173
recrosses Dan River, 177
Pyle's Hacking, 178-182
rejoins Greene's army,
183
skirmishes in NC, 187
commands party of
observation, 188-189

New Garden Road skirmish, 191-193
at battle of Guilford Courthouse, 193-201
account of 3rd line fight, 202
pursues Cornwallis south, 206-208
attack on Ft. Watson, 210-212
correspondence with Gen. Greene, 217-218
attacks Ft. Motte & Ft. Granby, 219-221
captures Ft. Galphin, 222-223
siege of Augusta, 237-234
screens Greene's retreat, 235-237
joins Marion and Sumter at Biggins Church, 239
skirmish at Quinby Bridge, 240-243
joins Greene, 243-244
battle of Eutaw Spring, 246-255
goes to Virginia, 256
engaged to Matilda Lee, 257
leaves the army, 257

correspondence with Gen. Greene, 258-265
life after the army, 267-270
death, 271

Lee, Lucy Grymes, 1
Lee, Matilda Ludwell, 256-257, 267, 269-270
Lee, Philip Ludwell, 256
Lee, Richard Henry, 1, 12, 15, 25
Lee, Richard Squire, 10, 13-14
Lee, Robert E., 270-271
Lee, William, 9-10, 17
Leesylvania, VA, 1, 8, 12-13, 22, 132
Lexington & Concord, MA, 21
Lincoln, Gen. Benjamin, 124
Lindsay, Lt. William, 66, 68-69, 77, 81
Long Island, Battle of, 48, 50
Loudoun County, VA, 25

Luzerne, Chevalier de la, 143
McCallister, Lt., 110-111

McKay, Lt. James, 211
McLane, Capt. Allen, 96-98, 100, 109, 111, 132-134, 139-140, 145
McMichael, Lt. James, 51
Madison Jr., James, 12
Maham, Maj. Hezekiah, 210-211
Malmedy, Col. Francois, 247-248
Manning, Lt., 252-253
Marion, Gen. Francis, 147-148, 150, 158, 208, 210, 212-213, 217-220, 232-233, 239, 242, 247-248
Marjoribanks, Maj. John, 251, 253
Marshall, John, 25, 76
Martin, Joseph Plum, 128
Mason, George, 16
Massachusetts, 19
Maxwell, Maj. Andrew, 220-221
Maxwell, General William, 44, 47
Maxwell's Light Infantry Corps, 43-45, 48
Middle Brook, NJ, 95
Miller, Capt. Patrick, 165-167
Monks Corner, SC, 239

Monmouth, battle of, 83-84
Monmouth County, NJ, 123, 126
Montresor, Capt. John, 44-45
Morgan, Col. Daniel, 61-62, 65-67, 196
   commands at Cowpens, 151-157
   retreats north, 158
   leaves army, 161
Morris, Col. Lewis, 260
Morristown, NJ, 30, 35-36, 38, 126, 134
   encampment, 1779-80, 127-132
Motte, Rebecca, 219
Motte's Plantation, SC 246
Mount Vernon, VA, 4, 20-21, 23
Moylan, Col. Stephen, 36, 92, 95
Muhlenberg, Gen. Peter, 115-116
Murfree, Maj. Hardy, 101
Murray, Governor John, Earl of Dunmore, 15-16, 21-22, 26

Nassau Hall,
  College of NJ, 11
New Jersey Gazette,
  67, 69
Newtown, Square, PA,
  67
Nicholas, Robert Carter,
  19
Ninety-Six, SC, 151, 210,
  235
  siege of, 227-234
Ninham, Capt. Abraham,
  87
Ninham, Daniel, 87
Nomini Hall, VA, 14
O'Hara, Gen., 159
Ogden, Col. Matthias, 138
Oldham, Capt., 221
Orangeburg, SC, 235-236,
  239
Pee Dee River, SC, 147-
  148, 161-162
Pendleton, Edmund, 19
Peyton, Henry, 25, 68, 77,
  81, 112, 133, 145
Philadelphia, PA, 15-16,
  20, 25, 29, 39-42, 46,
  54, 56-57, 60, 62, 65,
  67, 70, 82, 132, 136

Pickens, Gen. Andrew,
  177-179, 181-182, 221,
  223-224, 247-248
Poor, Gen. Enoch, 126
Powles Hook, NJ, 105,
  108, 264
  attack on, 109-114,
  117, 120-121
Preston, Col. William,
  183, 185-187
Prince William County,
  VA, 1-2, 4, 17
  militia in French &
  Indian War, 2-3
  Independent Militia
  Company, 17
  militia, 23-24
Princeton, Battle of, 30
Pyle, Col. John, 178-182
Pyle's Hacking, 178-182
Quebec, 140, 151
Queen's Rangers, 138
Radnor Meeting House,
  PA, 65-67
Ramsay's Mill, NC, 207-
  208
Randolph, Peyton,
  15

Rawdon, Lord Francis, 212-213, 216-219, 230-231, 233-236, 239, 244
Reed, Joseph, 116-117
Reedy Fork Creek, NC, 183, 185
Richmond, VA, 18
Richmond Country, VA, 1
Ross, Lt. John, 103
Rudolph, Michael,
  Lt., 90, 109-111
  Capt., 223, 247
  Maj., 268
Sandy Hook, NJ, 126
Salisbury, NC, 159, 161
Santee River, SC, 147-148, 208, 210, 218-219
Saratoga, Battle of, 61, 144, 151
Saumarez, Thomas, 198
Savannah, GA, 91, 124
Scammell, Maj. Alexander, 73
Scott, Gen. Charles, 84, 87-91
Scott's Farm, PA, 66, 73, 264
  raid on, 68-71
Seldon, Lt. Samuel, 233-234

Seymour, William, 214
Sheldon, Col. Elisha, 37, 92, 96
Shippen, William, 12
Shirley Plantation, VA, 270
Shubrick's Plantation, SC, 242
Simcoe, Lt. Col. John, 138
Smallpox, 38
Smallwood, Col., 29
Springfield, NJ
  battle of, 134-139, 264
St. Clair, John, 3
St. John's Church, Richmond, VA, 18
Stamp Act, 9
Stedman, Charles, 200, 202
Stephen, General Adam, 47, 50
Steuben, General Baron Von, 76, 132
Stevens, Gen. Edward, 196, 199, 203
Steward, Maj. John, 87
Stewart, Lt. Col. Alexander, 235-236, 244, 246, 248, 250-252

Stockbridge Indians, 87
Stuart, Lt. Col. James, 204
Stony Point, NY, 98
   battle of, 100-105
Stratford Hall, VA, 256, 267, 269
Sullivan, General John, 37, 47-48, 50
Sumner, Gen. Jethro, 247, 249
Sumter, Gen. Thomas, 213, 232-233
Sutherland, Maj. Nicholas, 111
Tarleton, Lt. Col. Banastre, 151-157, 165, 178-179, 182-183, 185-186, 192-193, 198, 201
Thacher, Dr. James, 129
Townshend Duties, 9, 13, 15
Treaty of Paris, 269
Trenton, NJ, 66-67
   battle of, 30
Tucker, Maj. St. George, 202-203
U.S. House of Representatives, 270
Valley Forge, PA, 62, 65-68, 92
   encampment, 73-76

Virginia Constitutional Ratification Convention, 270
Virginia Gazette,
   Purdie, 82
   Rind, 12,
Virginia Light Dragoons, 24-27, 29-34
Virginia Minute Battalions, 22
Wallace, Capt. Andrew, 155
War of 1812, 271
Washington, George, 1-2, 4, 9, 15, 20-21, 23, 27, 29-30, 35-40, 42-43, 46-48, 50-55, 57-59, 61, 65-66, 69-71, 73-74, 76-77, 79-80, 83-84, 86, 88, 91-93, 95-100, 105, 108, 110, 114, 116-117, 119, 123-124, 126-127, 130-131, 133-137, 139-144, 151, 188, 255-257, 262, 270
Washington, Lt. Col. William, 151, 153-157, 185, 188, 196, 199, 204-205, 214, 217, 232-233, 235, 248, 251-252, 262

Washington & Lee University, 271
Wateree River, SC, 219
Watson, Col. John, 218-219
Waxhaws, SC battle of, 151
Wayne, General Anthony, 47, 84, 100-101, 104-105, 118
Webster, Lt. Col. James, 185, 198, 203
Weedon, Gen. George, 114, 118, 127, 267
Weitzel's Mill, NC battle of, 184-187
Westmoreland County, VA, 1, 10, 13, 256
West Point, NY, 88, 98, 126, 136-137
Whiskey Rebellion, 270

White Plains, NY, 86, 88
Whitemarsh, PA, 60, 62
Williams, Col. Otho, 162, 164-167, 169-172, 178, 184-186, 197, 248-251, 253
Williamsburg, VA, 15, 23, 27
Wilmington, DE, 46
Winchester, VA, 2
Witherspoon, John, 12
Woodbury, NJ, 93-94
Woodford, Gen. William, 115-116
Yadkin River, NC, 159, 161
Yorktown, battle of, 255-257
Young, Thomas, 154

www.ingramcontent.com/pod-product-compliance
Lightning Source LLC
Chambersburg PA
CBHW060816190426
43197CB00038B/1766